FIRE
IN
THE FOREST

FIRE

IN THE FOREST

Dedicated to Those Who Have Fallen in the Fight

PHILIP SMITH

To order additional copies of this book, contact:
Xlibris
1-888-795-4274
www.Xlibris.com
Orders@Xlibris.com
745497

CONTENTS

A **fire lookout tower**, **fire tower** or **lookout tower**, provides housing and protection for a person known as a "fire lookout" whose duty it is to search for wildfires in the wilderness. The fire lookout tower is a small building, usually located on the summit of a mountain or other high vantage point, in order to maximize the viewing distance and range, known as *view shed*. From this vantage point the fire lookout can see any trace of smoke that may develop, determine the location by using a device known as an *Osborne Fire Finder*, and call fire suppression personnel to the fire.

The typical fire lookout tower consists of a small room, known as a *cab* located atop a large steel, or wooden tower. However, sometimes natural rock may be used to create a lower platform.

Mingus Mountain

Prescott NF Yavapai Co.

1911: Mingus Mountain was first used as a daily patrol lookout.

June 18, 1921: "Timbers have been cut for the Mingus Mountain fire lookout tower of the forest service which will be built immediately. It will be 44 feet high." (Prescott Evening Courier)

June 9, 1923: "It will be interesting to note that 10 minutes after the sounding of the siren, the fire guard at the Mingus Mountain lookout, reported the smoke to H. Basil Wales, supervisor of the Prescott national forest office, and said a fire was burning in South Prescott. The Mingus Mountain lookout is 22 miles from Prescott, across country. The guard rode a mule half a mile to a telephone and put in a long distance call through Jerome, and his report reached town only 10 minutes after the discovery of the fire here." (Prescott Evening Courier)

May 20, 1927: "V.J. Hopkins, formerly a mechanic in the Floyd Williams Motor company, arrived in Prescott yesterday morning from San Diego, Calif., to take over the duties of lookout in the tower atop Mingus mountain. Constant lookout, every day and Sunday, too, is maintained during the summer season at this tower because it is during this season that fires usually start.

July 13, 1932: "Forest Lookout William Anderson on Mingus Mountain had an experience with lightning a few days ago he swears he never will have to go through again. The incident was related today by Assistant Supervisor Llew J. Putsch of the Prescott national forest, who returned yesterday afternoon from a trip over at the Jaeger Canyon ranger station on Mingus. Anderson's lookout tower is about forty feet above the ground. The lookouts have instructions, when they see the approach of an electrical storm, to evacuate their tower and take refuge elsewhere. Well, the day Anderson had his big experience he saw just such a storm approaching and forthwith deserted his tower and skeedadled for his lookout cabin, only a short distance away. The lookout cabins, by the way, are a network of what might be called lightning rods in order to divert the electricity should it pick out the cabin or a place nearby to strike. Anderson told the assistant supervisor the approaching electric storm was not long in arriving and when it did get in his neighborhood it was in an awful reckless mood. Thunderbolts were cracking around everywhere and while it was happening, over a period of fifteen or twenty minutes, there stood Anderson in the middle of his cabin, as far away from the four walls as he could get. And good reason, too, because he said all about him the lightning was flashing down those wire conduits, just as though he were going to be put on the spot and all this preliminary performance was just to give him time to say his prayers. However, Anderson was let off that time but of course he had a few grey hairs for the experience, especially when not more than a hundred yards away from his cabin he discovered a large pine tree had been struck by a bolt and knocked into smithereens. Some people figure a lookout's job is a snap, but how about it now?" (Prescott Evening Courier)

February 10, 1944: "TWENTY YEARS AGO" "Three new lookout stations will be in service when fire season opens next summer, according to the supervisor of the Prescott national forest. One on Mingus Mountain, near the recreational area established on the summit last fall, is under construction a cabin and tower being built. Towers on Mt. Union and Tower Mountain have been completed but the apparatus for triangulation work in locating smoke and equipment for fighting flames have not been installed." (Prescott Evening Courier)

June 21, 1956: "A forest fire roared over Mingus Mountain in the Prescott National Forest, twenty miles east of here (Prescott), today, barely missing a number of summer camp buildings and a forest service lookout post.

June 22, 1956: "Firefighters Friday had most of the Mingus Mountain forest fire contained by fire lines and were working on a few hard-to-reach open spots. Ranger Oscar McClure, spokesman for the Forest Service here, said that while the fire has not yet been listed as under control, the situation is steadily improving and 'most of the line is being cleared up.' The fire has covered 13,500 acres, about one-fifth of it timber. It burned over the top of the 7,700 foot mountain peak. The fire broke out Tuesday when a range experiment being conducted by the University of Arizona just outside Prescott National Forest got out of hand. The flames raced up the brush covered mountainside into the timber. The fire at one time threatened a lookout station and a number of summer camps but the danger is now reported ended although a high wind could cause further trouble. About 600 firefighters, most of them trained Indian crews, are working on the fire. (Arizona Daily Sun).

*To apply for my old job

The Verde Ranger District, on the Prescott N.F. is offering a temporary Fire Detection position at Mingus Lookout Tower for the 2012 Fire Season. Fire Lookouts are often the single most important link to ground forces while they are engaged in firefighting, prescribed fires and general project work. Our lookouts must be experienced, reliable, attentive and most importantly, trustworthy as they may have a larger perspective of the fire environment, how incoming weather will impact it and the safety of personnel in the field. Our lookouts may be the only link our ground forces have to our emergency dispatch center. Duties as a fire detection lookout also encompass making public contacts and providing information to the public on the forest and Fire Prevention Measures. Lookouts spend time educating visitors about the U.S. Forest Service, the lands we serve, the wildfire environment, providing recreation information and sharing the beauty and awe of surrounding landscapes.

Recommendations:

Applicants are recommended to have a fire suppression background, in depth knowledge of Wildland Fire Behavior in a variety of vegetation types, possess the ability to accurately detect the location of a fire, give map reference/compass reading of the location using an Osborne Fire Finder and communicate the fire behavior, weather events and safety concerns to emergency dispatchers and firefighters. Intimate knowledge of natural weather patterns, weather conditions (wind events, lightning, dry conditions) and how it effects the fire environment is very important. Fire Detection Lookouts need to be skilled in basic/Intermediate radio operations. For a unique look into the fire detection world in Arizona, I recommend picking up the book by Eileen Moore, "Standing Watch, The Fire Towers of Arizona," for some light reading. It highlights the much loved Mingus Lookout Tower in which we are looking to staff this coming year. We look forward to your interest in this highly competitive position,

MINGUS MOUNTAIN FIRE LOOKOUT
BETWEEN JEROME AND PRESCOTT.

The heyday of fire lookout towers was from 1930 to 1950. During WWII the Aircraft Warning Service operated from mid-1941 to mid-1944. Fire lookouts were assigned duty as enemy aircraft spotters. The fire towers took a back seat to technology with aircraft and radio improvements.

Facts On Fire

FIRE IN THE UNITED STATES

The U.S. has one of the highest fire death rates in the industrialized world. For 1998, the U.S. fire death rate was 14.9 deaths per million population.

Between 1994 and 1998, an average of 4,400 Americans lost their lives and another 25,100 were injured annually as the result of fire.

About 100 firefighters are killed each year in duty-related incidents.

Each year, fire kills more Americans than all natural disasters combined.

Fire is the third leading cause of accidental death in the home; at least 80 percent of all fire deaths occur in residences.

About 2 million fires are reported each year. Many others go unreported, causing additional injuries and property loss.

Direct property loss due to fires is estimated at $8.6 billion annually.

Where Fires Occur

There were 1,755,000 fires in the United States in 1998. Of these:

41% were Outside Fires
29% were Structure Fires
22% were Vehicle Fires
8 % were fires of other types

These are called "hot strikes" due to the visible thickness of the lightning. Sometimes, when a large storm is crossing a lookouts' area, they will stay up for the duration of the storm. The visible large strikes are recorded by degree. The next morning when the burning period begins (10a.m. to noon),

The recorded angle from the previous night is checked to make sure no fire has begun.

The aftermath of a large fire in California. This conflagration roared through Mt. Wilson in the Los Angeles area. Destruction of almost all brush and grasses is evident. The lone survivors are the Yucca plants seen in the foreground adjacent to the highway.

Wildland Firefighting Tools

1. McLeod: a tool with a large blade resembling a very large hoe on one side, the other has a blade with large, separated teeth. Used in the high country for raking fire lines with the teeth side and breaking up hard sod and cutting branches with the hoe side. Very useful for scraping a berm line or digging through layers of pine duff. It is a bothersome tool to carry around, making sure the toothy part is away from your body. Duct tape usually covers the cutting edges when the tool is not in use. When in use it is carried with the tines pointing down. When you are resting on a fire line, stand the tool on its head instead of flat on the ground. In Australia, it is called a "Rake Hoe". Named after ranger Malcolm Mcleod.

2. Pulaski: A combination of an axe and adze in one head, the handles are made of wood, plastic or fiberglass. Basically, the axe is used for chopping wood, the adze for breaking up hard soil. Not to be used in rocky soils, as the axe part is easily breakable.

    In a disastrous fire in August of 1910, one forest ranger (Ed Pulaski) saw the need for a better firefighting tool. By 1913, Ed's invention was in wide use. By 1920, the forest service contracted for the tool with commercial manufactures, In todays' 2012 fire tool catalogue, they sell for about $70.00. In early times, in the 1950's, a forest district had its own colors painted on the handles. My district, the Mesa, part of the Tonto National Forest, colors were red and black. My hard hat was also painted in a similar fashion. At one time, on any distant fire, we could tell our own district's fire crew members by seeing our colors on the fire tools.

3. Mattock: similar to a Mcleod. It has an adze blade that functions as a hoe in hard ground, the other blade can be a pick for prying up small rocks or a cutting edge for chopping roots. They are usually 3' in length and the head weights can vary from 3 to 6 pounds. Keeping the head tight in use can be a chore. Small sheet metal screws are often placed just below the head to secure it on the handle. The handle is removable when packing.

4. Hoes: Various weights are available. Useful for repairing trail tread and digging trenches to hold waterbars. Usually the handle is 34" long with a six inch wide blade set at an angle. Used like a mattock. Grub hoes are not usually sharpened, and have a removable handle for ease in packing.

5. Pick (Pick-axe): Has a chisel blade on one end and pointed tip on the other. Useful for breaking up heavy soils and gravel, or digging holes/trenches. Do not use to pry loose large rocks. An awkward tool in some ways, and a day swinging these will guarantee a good night's rest.

6. Flapper: aka(Swatter or Beater). Strips of heavy material called lamellas are attached to a long handled frame. Useful in certain areas and conditions.

Can be pulled along a slow moving grass fire and work very well. If used too vigorously, the downdraft created can actually spread a fire.

Only used in very minor flareups. Developed from the idea of using wet sacks with which to "swat" a fire, known as "wet sacking".

I used them once on a small grass fire around Sunflower, Arizona. They were unwieldy and didn't work for hell on that small burn. We swapped them quickly for piss bags and had better success as the wind was about zero. "Piss Bags" are rubber containers with shoulder straps, that hold several gallons of water, they have a pump like apparatus and variable nozzle. They work great under certain conditions. They were great tools for knocking down fires running up the trunks of giant ponderosas and spruce on a fire by Yosemite National Park (actually on the Stanislaus District). The amount of water contained is used up fast, and initially when full, they are heavy and awkward to use.

A Fireman's gear through time

World War II gear, such as belts became standard. Military terms as "attack" were absorbed into the forest service fire fighting vocabulary. In the 1940's, canteens were secured on your back. The earlier web belts and canteens were metal. Your name and serial number were scratched into them. They were cleaned weekly by adding one tablespoon of baking soda.

Canteens were changed from metal to plastic due to the fear of disease. Newer canteens were two quart capacity to which instant tea was added as it tasted good whether warm or cold.

A third punch held head lamps, with the wire under your undershirt. Hard hats had two attachments, two in the front, one in the back. Headlamps had varying intensity. They are sophisticated now with LED'S. An updated fire shirt is worn ($39.00). Plastic goggles were added attached to clips, and safety glasses are also used. The one essential was the fire shelter changing shape through time. These were attached to your web belt in the 70's, and their efficiency greatly improved through the decades (protection from higher and higher heat numbers). Everyone practiced the technique of unwrapping the shelter and getting in putting your feet and hands in the attached interior webbing. I think ninety seconds was allowed to pass this test.

A scabbard holding file and handle are added. A blown out part of an old fire hose with rivets are now added to the belt. (which held the file and handle). Portable food came in square packs with a size identical to C rats (C rations - also courtesy of WWII). Because of the Vietnam War, MRE's(meals ready to eat joined the list). These were self- heating by some mysterious chemical reaction.

Chunky soup was added, mac and cheese were a favorite. People would swap anything for franks 'n beans. And last was added a jacket type made from old parachute shrouds to secure it to the body. Harnesses for extra gear were added, as the belt grommets held even more. Add a radio (only those in the middle or higher in the pecking order could have one).

In a twenty man crew, it is common to use three radios. One at each end and one in the middle in case the crews get widely spread. Vests were added like fishing vests with many smaller pockets. First aid kits were on the list, another item on the vertical seniority list, only squad bosses have these.

A man in the middle of the fire crew was selected. The first aid person cannot be tied up with cutting tree limbs and scraping line. Tube pockets allowed even more stuff. Pens, pencils, toothpaste, tooth brushes, beef jerky, cigars, but mainly the now known to be lethal snuff- Copenhagen, Skoal, Beechnut. The fire line person now is carrying about 35 pounds. In the fall in cooler weather, extra shirts, small jackets, especially on the night shifts. The fire shelter also served as a pillow on overnight fires, and has grommets for a shoulder harness, as above, aka fire pillow.

Shrinks tell us that it takes two weeks for the body/brain to readjust from a 24 hour fire shift.

Two long fusees can be added as backfire torches. If it all goes to hell, bring out the fusees and burn out the fuel between you and the fire.

Not to forget - in ancient times, the buck knife. Positioned on a regular pants belt. The belt buckle has requirements, the larger the more powerful. power words as Bud, CSA, American flag, also noted are naked ladies and Mack Trucks. Old time buck knifes were made of Solingen steel (Germany), the first were formed and shaped from a handmade rotary lawnmower blade.

Newer bucks have far less metal quality. Recently, the buck knife has been replaced by first the Swiss Army Knife and now the Leatherman.

Enter the electronic age. Ipods, Ipads, laptops, cell and smart phones can all be found in the newer packs.

Rookie fireman joining the crews midseason would buy a J.C. Penny 8" boot with questionable sole material. Regulars wear a 10" smokejumper boot #375, black leather with a piece of rigid steel. Made from timber brown elk. Anyone not wearing 10" Whites (most recently $399.90) was considered to be an inferior. The White's last almost forever. Protect your ankles like a second coat of stiff skin. Return'em when you burn'em out to be resoled or whatever else is needed. Still made in California. One newbie wearing cheap boots had to walk through a burned area. As he walked the soles of his boot were literally being consumed, layer by layer. By the time he exited the fire area, the sole had melted from the ash and left only foam. Someone took their knife and trimmed his soles.

Getting to a fire cross country allowed the use of a Lockheed Electra. Holding 80 people (4-20 person crews). On one occasion, 80 buck knives were confiscated prior to take off. God knows how they returned each man his own knife, like returning .45 Colts from a western bar.

When vital supplies as water run low, choppers deliver supplies to a designated drop area or local helipad if one is close. Five gallon collapsible cubicles for water are dropped as close as possible along with other rations.

One crew from a church related group called themselves the Shasta Trinity (Seekers: Matt 7:3). Seems the Seekers were serious about their faith. They held an impromptu service, which one of the Little T men visited. The Seeker-speaker was the Preacher from then on. Even more strange, the little T crewman caught the spirit. He asked if he could speak to the carnal, Copenhagen chewing crowd. they were deeply moved. Hardhats off, listening to the regular unseminary trained crewman, it became part of their schedule.

Some of the fire crews represent their individual reservations or tribes. They have a proud history of fire fighting in both the national forests and on reservation lands. Some of the highly respected tribal hot shots include, Fort Apache, White River Arizona, Warm Springs IAG, Navajo, Golden Eagle, Geronimo (San Carlos, Arizona), Zuni and Chief Mountain (Blackfoot Nation). Additional and prolific information is available from Jim Steele – IAG. (406-676-2550_

Fire tools (l. to r.)
1. Scott (lady shovel).
2. Russ Copp (brush hook).
3. Spitter (Mcleod).
4. Tom Tillman (Pulaski).

This is about 1972. Location is the Sunflower Work Center on the Beeline Highway between Phoenix and Payson, Arizona.

Fire Fighting Training and Activities - Conditioning:

In the morning physical training ensues. Sunflower followed the Canadian military example. Climbing of a very long rope attached to a tree limb (the ancient sycamores in Sunflower's case), many push and sit ups, a four mile modified run with a forty pound field pack and some stretching were the basics.

Tools:

After a fire the tools are seriously degraded. Hands on sharpening of Mcleods, Pulaskis, axes and shovels (several kinds) is a priority. Secured in a heavy vise, the handles are well wrapped with heavy material to secure same. Filing is done with heavy gloves and bastard files. Edges are hand filed to a razor sharpness then made safe by using duct tape for both safety and storage. Chain saws likewise are taken apart and filters replaced, oil reservoirs and gas refilled, and the chains sharpened manually. This is very time consuming.

Painting:

And you thought the Navy was bad. If it moves we throw it out, if it's anchored we paint it. Everything. Stairs, outer and inner walls, doors, furniture. Most of the barracks are 40 or 50 models and are very dry, so much paint is needed. Basic forest service green was the choice of color.

Inventory:

A nonstop event. Gasoline, spare tires, spare tools, water, c-rats(rations)-this was before the MRE's (meals ready to eat) were available. At the Sunflower Work Center, my food supply was the C-rats, some from WWII. and still very edible. However, the Spanish Rice was considered lethal and chosen by no one. Also, fire shelters, canteens, hardhats, rope and so on und so weiter.

Terrain Improvement:

A misnomer for activities occurring during "down time". These included raking the grounds, cutting limbs from brush, low hanging branches (no limb should be low enough to knock your hat off),

Stream bed small diversion dams, some 3 or 4 feet deep (in the summers of Arizona, these were a welcome relief). Erosion prevention, small berms, rock filled heavy wire (aka gabions) holding back steep runoff areas, fire tower step replacements and anything else sorely needed by people back in their air conditioned offices. Sometimes a new helipad was needed on a mountain top, sometimes with flat pushed dirt areas, sometimes with concrete poured landing slabs. Any trees, shrubs, rocks that would interfere with a choppers descent were removed. These added to the transportation of fire crews during the fire season.

Meals:

An army may travel on its' stomach, but so did the fire fighters. Most meals were cooked, usually by someone chosen for that ability, but not always. Needless to say, tofu, squid, most vegetables and lite oatmeal were not seen. Heavy on the spaghetti, dead mammals of any kind, fried anything and steak when possible were the primary choices. Many barracks also had a nearby restaurant for Friday evenings. Sunflower had the Sunflower Store. Lunch was often sandwiches. One person would take a variable desired list and lots of cash and return with the meals, minus the cash.

Eating on a fire line is another matter. If you're in the outback, whatever you have in your backpack is what you're going to have. On some occasions near semi-civilized roads, a cold KFC series of meals would arrive, delivered by some ancient truck usually in the early morning hours. Really appreciated even when cold. It can take a long time to locate a fire crew in the middle of the night, even when the roads are familiar.

Horses:

The equines bought by the F. S. seemed to be the very nervous, very aged, knock-kneed types no one else would have. Bought at remote auctions for very little cash. Sunflower had a small remuda of these. Even city slickers knew you didn't want to ride on a horse named Billy the Kid, Preacher or High Pockets (he was on the Globe District and much loved by former ranger Wild Bill Buck.) One of sunflowers elite nags would refuse to cross the tiniest of streams. One had to dismount and lead him or her gently across, then all was well. Another cayuse would come to a split in the trail. You wanted to go right, he/she went left, and after considerable trail time, one could turn around go back and finally make the right trail.

We had a mule named Killer with which me, Dave Slan, Bill Blackwell and J. R. Parmier built a couple miles of fence high up under Four Peaks. We were about 6,000' up working pounders in DG (decomposed granite). It took everyone in sight to back said mule out of the trailer, and not gently.

Once loaded with two rolls of wire on both sides of the X framed wooden pack saddle, I would then lead the gentle animal parallel to the fence line. Killer would note in his mind that the next hill in front of us was getting close, then he would light for the high country-we would get out of the way. Once he fell with a crash just below the crest of the hill. His neck was bent back and I said, "Jeez! guys, he's dead. However, as soon as we removed the wire he was happy as hell again, and we would start all over. (See Killer chapter).

With the wire stretchers and Killer pulling the wires out one at a time while the wire roll pivoted was a neat thing. Lots of "Dead Men" were added. (These are heavy stones, wire wrapped and attached so there are no escape holes for cattle, then secured to the wires above).

We put up lots of wire, drove lots of metal T stakes with the pounders. We didn't finish the whole job. It got so cold in the high country we thought the mule might die, as we left him in a corral each evening. We left him to graze and think happy thoughts at the Blue Point Station on the Verde River. Somehow he wasn't happy and lit for the reservation. Somebody said they'd seen him dancing with some other horses and/or mules along the Verde River. I never knew his ending.

Standard Firefighting Orders and 18 Watchout Situations

The original ten Standard Firefighting Orders were developed in 1957 by a task force commissioned by the USDA-Forest Service Chief Richard E. McArdle. The task force reviewed the records of 16 tragedy fires that occurred from 1937 to 1956. The Standard Firefighting Orders were based in part on the successful "General Orders" used by the United States Armed Forces. The Standard Firefighting Orders are organized in a deliberate and sequential way to be implemented systematically and applied to all fire situations.

STANDARD TEN FIRE FIGHTING ORDERS

1. Keep informed on fire weather conditions and forecasts.
2. Know what your fire is doing at all times.
3. Base all actions on current and expected behavior.
4. Identify escape routes and safety zones and let them be known.
5. Post lookouts when there is possible danger.
6. Be Alert. Stay calm. Think clearly. Act decisively.
7. Maintain strong communications with your forces, supervisors and others involved.
8. Give clear instructions and make sure they are understood.

9. Maintain control of your forces at all times.
10. Fight fire aggressively, keeping safety first

Wildfire crew information updated

Source: Fire staff/ Mesa Ranger District/ Tonto National Forest/ February, 2012,

Not all districts on the Tonto have crews living in barracks during the fire season. Only Mesa, Pleasant Valley and Tonto Basin.

None of the districts provide meal services anymore.

Fires are allowed to burn as part of the natural process. All fires that threaten homes or public infrastructures are fought.

Most fire crews are of the same size. The Tonto has 3-20 person hot shot crews, 1-10 person Helitack crew and 15 engines with 5 to 8 people (depending on funding). Also 3 Prevention Techs.

Firefighters need to be qualified continually. A task book must also be completed before they are certified. (We used to get red cards, which meant we were allowed on a fire line).

Weather satellites are the only imagery the currently uses. There is a lightning detection system and gives approximate fire locations, plus the number of lightning strikes. Google earth is also used.

Makeup of crews is determined by demographic census data regarding the breakdown of crews as to race, sex, age and background.

Angeles National Forest History

On December 20, 1892, the San Gabriel Timberland Reserve was created by President Harrison. The creation of the Reserve, which was the forerunner of the Angeles, was in response to public concern about watershed values as early as 1883. Floods resulting from fire denuded slopes were causing problems with the lowland populations. In 1905, the Reserves were transferred from the Department of the Interior to the Department of Agriculture, and renamed National Forest in 1907. The San Gabriel National Forest consisted of the southern section of the present day Angeles and portions of the San Bernardino Forest. In 1908, the name was changed to Angeles National Forest.

Above Photo: one of the lookout towers on the Angeles National Forest. Note how smoggy the view is to the lookout. Many formerly utilized towers are locked up due to the fact the air quality is so poor.

Station wildfire

More than 161,000 acres (650 km²) of the forest were burned by an arson fire that began on August 26, 2009, near Angeles Crest Highway in La Cañada and quickly spread, fueled by dry brush that had not burned for over 150 years. The fire burned for more than a month and was the worst in Los Angeles County history, charring one-fourth of the forest (250 square miles), displacing wildlife, and destroying 91 homes, cabins and outbuildings and the family-owned Hidden Springs cafe. During the fire, two firefighters died after driving off the Mt.Gleason County Road looking for an alternate route to get the inmates out at Camp 16.

The "Station Fire" threatened the Mount Wilson Observatory atop Mt. Wilson. The site includes two telescopes, two solar towers, and transmitters for 22 television stations, several FM radio stations, and police and fire department emergency channels.

Phil (from Chet Ogan, Mill Creek Fireman, email (March 3, 2012).

You may have met Bobby Olsen once. In 1973 he was the patrolman stationed at Soledad Guard Station down on the old highway through Soledad Canyon between Placerita and Palmdale. Gary and I worked there a couple of days rewiring the station house. Bobby was there with his new wife Arlene. I think the rest of the crew did cleanup and fix up work

Bobby was lead cutter on the Little T Hotshots in 1970 to 1972. He and I were two older experienced crewmen hired on the crew. Bobby set the pace for the crew, he was a fast cutter and a very hard worker. He set a fast pace.

Bobby was a Vietnam Veteran, was in forward areas where Agent Orange was liberally applied to foliage. He continued his career with the Forest Service until he retired about 10 years ago on the Lassen NF as a fuels management specialist. He conceived the idea of the Ronald McDonald House program for underprivileged children.

About 6 years ago he developed symptoms of diabetes caused by exposure to Agent Orange and eventually lost a leg. Bobby passed away last month, there will be a remembrance of his life in Susanville in June.

Back row: George Scribner (see: Tribute chapter).
Middle row:?, Chet Ogan, Gary Reynolds,?, Stevenson.
Front row: Smoke jumper from Idaho,?, Ron Emeterio.
Phil Smith not in photo, as he's always the photographer.

Fri, July 15, 2011
Hi Phil,

There's nothing to rebuild Mill Creek for anymore. Pretty much scorched earth when I drove thru last October. Apache Saddle has been downgraded to just a patrolman station. Most of the activity for the district is out of Bear Divide and Little Tujunga which were unscathed. The district office is at Little T now, moved out of the Sylmar location. Little T has been operated as a regional training center since the hotshot crew left about 1984. The hotshot crew has arisen like a phoenix and is now operating again. A nice bunch of kids. The complexion has totally changed.When I was on the crew in 1970 we were racially mixed- mostly white with 3 or 4 Hispanics, one Japanese, one Negro, age mixture from 18 to 30 years old. Now the crew is 70% Hispanic. They have a couple of exchange firefighters there. Three crewmembers from BellaRussia served for two years and last year four from Australia.
Source: Chet Ogan, former Mill Creek Firecrew member.

Station wildfire

More than 161,000 acres (650 km²) of the forest were burned by an arson fire that began on August 26, 2009, near Angeles Crest Highway in La Cañada and quickly spread, fueled by dry brush that had not burned for over 150 years. The fire burned for more than a month and was the worst in Los Angeles County history, charring one-fourth of the forest (250 square miles), displacing wildlife, and destroying 91 homes, cabins and outbuildings and the family-owned Hidden Springs cafe. During the fire, two firefighters died after driving off the Mt.Gleason County Road looking for an alternate route to get the inmates out at Camp 16.

The "Station Fire" threatened the Mount Wilson Observatory atop Mt. Wilson. The site includes two telescopes, two solar towers, and transmitters for 22 television stations, several FM radio stations, and police and fire department emergency channels.

Sun, February 12, 2012 12:49:47 PM

Phil,

We cooked our own meals. That year we met with Bob Brady, a qualified tree faller near Mt Gleason to fell some incense trees which we split into rails to put up a fence beside the parking areas for the crew parking and public parking at the station. In some of the long splits we used all the metal wedges we had, I cut some long wedges out of some stout oak limbs that could reach clear thru the cedar cants.

That trout place was Hidden Springs, right next to Monte Cristo Station. That area burned in the Middle Fire in 1977. In the floods the following winter Monte Cristo Station foreman George Scribner was swept away with his truck trying to get across the creek there. I don't think he was ever found. The whole eastern part district burned again in the 2008 Station Fire; the southwestern part of the district burned the year previous. Major fire frequency there is about 30 years. Chaparral senesces at about 20 years- that is, there is more standing dead material in the plants than live material.

I recall doing something at a plantation up the hill above the station. We ran that section of road for PT each morning.

I was in pretty good shape. I remember doing the step-test then. Gary Helsel was checking my pulse after the required 15 or 30 seconds or whatever time interval it was. By using mind-over-matter I controlled my pulse while he was feeling it and dropped it from about 100 bpm to 60 bpm in about 5 seconds.

That must have been the year that I did a 3-day 27 mile hike in the Sierra Nevada near Piute Pass- probably Labor Day weekend with my roommate from the Valley- Dave Couch. Dave was running about 5 miles a day, so we were both in good shape, cardio-wise. Dave and I hiked in a couple of miles from North Lake (near Lake Sabrina) on a Friday night full moon after driving from LA. We laid our bags on the trail to sleep. It was cool in the morning. While still in my bag I started some hot water on my Svea stove for coffee and oatmeal. I think I might have eaten the oatmeal while still in the bag. We hiked up through Lamarck Col (12,830 ft) mostly cross-country because we missed a faint trail marked with rock cairns. As we crossed the col we looked back and saw the trace. We went into upper Darwin Canyon (near Evolution Valley) and camped at middle Darwin Lake the 2nd night. Still no mapped trail, just following the map and traversable terrain. We topped out at Alpine Col at 12,390 feet and climbed nearby Muriel Peak- 12,940 ft. The walk down the next valley was rugged. The rocks were the side of Volkswagens, we were either jumping from rock to rock or having to pass down our packs and crawl down between the boulders hand-carrying our packs because the gaps were too narrow for both us and our packs. We camped near lower Goethe Lake- 12,130 feet. The alpine-glow was beautiful. Another mile or so took us to the Humphrey's Basin trail at Piute Pass, our first trail since Saturday morning, and 9 miles back down to my Datsun 510. Of the 27 miles, only 9 miles was on trail. Neat trip! [I had to look at a topo map to recall the exact location].

Ah memories!!

When we try to pick out anything by itself, we find it hitched to everything else in the universe.

-John Muir

Chet Ogan
Eureka, CA

Chet Ogan famed biologist and forest service fire fighter on the Angeles National Forest.

We served on a tanker crew at Mill Creek Station, which burned to the ground and since been rebuilt.

Granite Fire

Three of us fire fighters sat in an open truck all night heading north from Los Angeles, going to the "big one". From about fifty miles away, we could see the huge glowing red glare, high up in the pine country. We arrived in the dead of night, but it seemed like high noon. When we took some back roads to get to the fire, it was dark again. Surrealistic shapes of bombed out giant trees with smoke coiling around, then no visibility due to the smoke and the heavy, acrid smell of smoke, we all craned for a better look. We were driving about two miles an hour, rounding curves in slow motion, looking through the smoke, not knowing where the head of the fire was located. Suddenly, we topped the smoke and felt better. Clean air. No smoke.

Found out later that this fire, which burned thousands of acres of prime douglas firs, coulter pines and ponderosas had been started deliberately, some young males had roared through a campground throwing out something on fire, the end result was hell.

We joined hundreds of others in fire camp and hit the smoke in the morning. From a water tanker, which was our only water supply, hoses are stretched out every one hundred feet, to which we attach, as many additional one hundred foot sections as needed. Once, I figured we had eleven thousand feet of one inch hose stretching out on all sides from the mother tanker. Our water supply lasted about three minutes with all hoses in use.

The white ash under fallen, giant trees would be two feet deep, sometimes cold, sometimes not. Standing in the ash would tell you. When your feet got too hot through the thick boots to stand it, you turned the hose on your feet and kept going. The fire movies just don't catch the out-of-doors feel.

In the movies, everyone hears all of the fire orders crisp and enunciated.

On a real fire, at times you can't hear yourself think. Wind ripping through the trees, Truck engines running, choppers directly overhead dropping water, Sawyers dropping trees right and left, sometimes too damn close, a sixty foot pine falling without a few yards gets one's attention.

The front going part of a fire is never attacked, you work from the back and both sides to try and turn the fire and gradually reduce its' size. That's the theory, but if the humidity stays high and the winds increase and the fuel is constant and the fire God is upset, or some chopper hotshot drops in for a look just as you have an area controlled, the thrill begins all over again.

I recall manning (peopleing to be politically correct?) a hose with my partner about thirty feet above me and one hundred feet away on the next hoselay, we put water down all day on the edge of a fire, throughout the day, the smoke would roll in, roll out, roll in and at times, he wasn't visible. He was just high enough that the smoke was thicker and he spent two days with eyes completely closed from the smoke, even though we both had two wet, giant red bandanas, soaked in water, one around the mouth and nose like a bank robber and the other covering the forehead under the hat. This created a small slit to see through, which was enough.

The Granite Fire cost millions and I would like to go back and see the recovery of the land. Maybe find the buck knife I dropped, or see the picnic tables by Cherry Dam again.

What's in a name? Granite fire? Naming fires after individuals is not allowed due to possible legal ramifications. Some fire names enter into history, fire folklore, etc; and often take on a life of their own. The stories get hashed and rehashed and with repeated telling become larger, more complex, add a measure of one-upmanship to the person telling the tale. The, "I was there, you weren't idea prevails In local wildland communities, having made a bunch of "major ragers" elevates a fireman into an elite group – along with bragging rights.

Having named such a conflagration myself, I know about bragging rights. The "Mattress Fire" along Hwy 88 is little known. The inferno was phoned in by somebody enroute to Phoenix from a pay phone. When the Sunflower crew arrived, it was already cold, a smoldering sea of smoke. The "mattess" which some hirsuted type had thrown off of his/her truck, had somehow caught fire. I think we stretched it's size to a ¼ acre. I also think we left the remnants. In our reflections on fires that year, somehow the "Mattress Fire" didn't enter the conversation.

The Pestigo and Chicago fires happened on the same day. Hundreds died at Pestigo plus numerous towns and villages, plus thousands of acres of valuable timberland were destroyed. But Mrs.' O'Leary's cow survived in legend.

When names like Loop, Storm King, Blackwater, Mann Gulch, Griffith Park, Canyon Inn, Rattlesnake or Inaja were chosen – could anyone have predicted the loss of life that occurred.

Based on their sheer size alone, the following reads like a who's who among major fires. Station, Marble-cone, Sundance, Scarface, Tillamook, Matillija, Sleeping Child, Wenatchee, Liebre, Bear, Laguna, Bel Air, etc;-some so large that they were called the Yellowstone Complex or the great Idaho Fire of 1910.

The author in the 1970's on the Granite Fire (Stanislaus N. F.).

Me and scores of others operated hoses - calming down a fire one stump at a time. Note the lack of Nomex gloves to protect the hands. I improvised two dampened bandanas, one above my eyes, the other on the top of my nose. Created a little slit in which one could breathe during the smokier moments. My fire shirt is long sleeved, which is a violation of ethical policy. Mine should've been rolled up. The only dudes who wear'em full length were the hotshot crews.

You will work in this dense smoke all day long keeping control of a 100' length of water bearing hose. Which will run empty very quickly.

It is about 1973 and the author is taking the photo.Because of a lack of eye protective gear, the barely seen smoker eater will not be able to see at all come the morrow.Not to forget the long term effects on the human lungs. But no one complains.

The fire must be out. The author, two guys I can't remember, and Prof. Chet Ogan, famed biologist from northern California. We get to go back to Mill Creek Station, and wait for the next "big one".

Indian Use Of Fire

Hunting - The burning of large areas was useful to divert big game (deer, elk, bison) into small unburned areas for easier hunting and provide open prairies/meadows (rather than brush and tall trees) where animals (including ducks and geese) like to dine on fresh, new grass sprouts. Fire was also used to drive game into impoundments, narrow chutes, into rivers or lakes, or over cliffs where the animals could be killed easily.

Crop management - Burning was used to harvest crops, especially tarweed, yucca, greens, and grass seed collection. In addition, fire was used to prevent abandoned fields from growing over and to clear areas for planting corn and tobacco Clearing ground of grass and brush to facilitate the gathering of acorns. Fire used to roast mescal and obtain salt from grasses.

Improve growth and yields - Fire was often used to improve grass for big game grazing (deer, elk, antelope, bison), horse pasturage, camas reproduction, seed plants, berry plants (especially raspberries, strawberries, and huckleberries), and tobacco.

Fireproof areas - Some indications that fire was used to protect certain medicine plants by clearing an area around the plants, as well as to fireproof areas, especially around settlements, from destructive wildfires. Fire was also used to keep prairies open from encroaching shrubs and trees.

Insect collection - Some tribes used a "fire surround" to collect & roast crickets, grasshoppers, pandora moths in pine forests, and collect honey from bees.

Pest management - Burning was sometimes used to reduce insects (black flies & mosquitos) and rodents, as well as kill mistletoe that invaded mesquite and oak trees and kill the tree moss favored by deer (thus forcing them to the valleys where hunting was easier).

Warfare & signaling - Use of fire to deprive the enemy of hiding places in tall grasses and underbrush in the woods for defense, as well as using fire for offensive reasons or to escape from their enemies. Smoke signals used to alert tribes about possible enemies or in gathering forces to combat enemies. Large fires also set to signal a gathering of tribes.

Clearing areas for travel - Fires were sometimes started to clear trails for travel through areas that were overgrown with grass or brush. Burned areas helped with providing better visibility through forests and brush lands for hunting and warfare purposes.

The author and Killer, famed mule carrier of large rolls of barbed wire (carefully loaded). At 7,000' altitude just this side (west) of the well known Four Peaks Mountains visible from Phoenix. The crew and I handbuilt almost two miles of range divider fence until the weather drove us out. The crew was fine, we were worried about the psycopath with four legs catching newmonia - we left him in a pathetic, small corral overnight located at what we laughingly called a "road," and resumed the fun each morning.

We returned Killer at his/her home base known as Blue Point on the Rio Verde. She/he didn't stay long. Took off for the local Pima res. Last we heard, she or he was cavorting with one of those young fillies on the reservation. Never heard from Killer again. She/he crossed over to the lush, green grasses of eternity where she/he no longer has to carry heavy rolls of really sharp barbed wire. Rest in peace - Killer. (See the chapter on "Killer.")

GRANITE FIRE - BY YOSEMITE - 1973. "Backfiring with a drip torch."

The premise is to meet fire with fire and to remove fuel from the oncoming blaze.

Usually an intersecting road or fuel break is chosen to begin the work. No one likes to see beautiful pines consumed, but the trade-off is wise. You hope to save an even larger area from the flames

Gary Helsel and others are doing the work.

The surrounding air can be very hot and swirling as the fighter tries to connect and lay out more hose. He hides behind the trees for protection. At times, thousands of feet of hose are interconnected to reach proximity to the advancing fire. These men are lucky in that there is no thick continuous fuel between the trees.

A small brushfire down in the canyons of L. A. County. USFS and LA firemen responded, as it was on the "redline", which means both fire agencies share duty.

About 1972, men on the lower right are checking for any small spot fires which may have started blowing over on small embers from the main fire.

It's a hundred degrees frying pan hot and your boots keep slippin' on the slope, the decomposed granite doesn't help. You're also dragging a hose over the ridge, so the lead man on the nozzle can open up the R3 and knock down the base of this small part of the roaring fire. You will do this all day. The mother tanker will deliver water all day for you.

Typical of the terrain and severe slopes on the Angeles National Forest just above Pasadena, California.

The Tonto National Forest

In 1905, the Tonto National Forest was created to protect the watersheds around reservoirs. Six major reservoirs could store more than 2 million acre-feet of water. The watersheds, riparian areas and water quality are all radically affected by major fires in the forest. Two major rivers, the Salt and the Verde provide refuge from the long season of heat, and in addition six man made water reservoirs are sources for camping, fishing, and just plain relaxing.

During the 1880's and 1890's the countless numbers of cattle brought in by westward migrating Texans and others, grazed the historic grasses to the point of extinction. Grasses that grew as high as a horse's belly were burned off in the Texas tradition, killing the roots. The end result was erosion and scrubs replacing the former grasses. To control the government land, the Tonto National Forest was created.

Beginning with the earliest dam, Roosevelt was 90% completed in April of 1910. It was finished and dedicated by President Teddy Roosevelt on March 18th, 1911. Many of his "rough riders" were in the crowd. Roosevelt was the fifth and final speaker of the day. His idea was to create a national irrigation project similar to the Panama Canal, and he hoped people would come add see this dam as they did Yosemite and the Grand Canyon. It was only 20 years since the Apache Wars were over, and many Apaches were used to build the roads adjoining the dam. The modern gawd-awful chuckholed, twisty and better be driving a very good 4WD truck trail leaves from Roosevelt Dam winds around the series of SRP (Salt River Project) dams and heads for the connecting highway between Phoenix and Payson.

Each national forest has separate districts. The Tonto National Forests has several usually named after the closest community it serves, so Cave Creek, Globe, Mesa, Payson, Pleasant Valley, Tonto Basin are all districts linked to the home office in Phoenix, Arizona.

There is a forest archeologist on each forest, plus a small staff of specialists. Their mission is a daunting one, with literally thousands of pre and historic archeological sites, ranging from the archaic thousands of years old, the cultures of Sinagua, Hohokam, Salado (prehistoric) and historic sites as various military forts (Reno, Camp Lincoln, Verde, McDowell,Pinal, Apache, and others. Thousands of sites are protected through several federal and state laws. There are site stewards assigned, people who live in a community and close to larger, largely prehistoric sites, to make sure that no illegal digging or collecting of any kind occurs. They serve at no cost to the agencies and their work lies in protecting the ancient sites for future analysis or simply protection. They are a serious, driven group of people.

The author on the left about 1972 with Les Chapman, a lifer in the forest service. Les and his wife Bertha lived in the little 1930's type house adjacent to the Sunflower Work Center. The house has a large basement into which they had placed their accumulations of a lifetime.

Came the rare giant flood roaring down Ash Creek filling the basement destroying most of their possessions and history. Continuing down the creek, the water also pushed in the north side of the 2 story barracks cleaning out everything inside the building. Chain saws were found unopened a mile away from the site.

The Chapmans were longtime fire lookouts on Mt. Ord. (I subbed for Bertha).

Les knew where the secret apple orchards down the Matazel trail (pronounced Mat- a- zels).

To save money many of the old signs were repainted (as is seen). Les was trapped on the wrong side of a roaring creek and spent two days in a small cave. None the worse for wear.

I visited him once in Punkin Center. Directions were typical western. Go to Jip Toots Store, cross Tonto Creek (if you can), head towards the tall cottonwoods about a mile away, small house with a rock chimney. About 111 degrees. Desert dwellers don't talk about real hot until it's 116 or so.

They were typical of the older tough breed of forest service workers. Both gone now.

Outdoor lovers, hard physical workers, never bitched, civil, well-mannered people. I miss 'em.

L

A brief history of the Signal Peak Fire Lookout area

The Pinal Mountains rise to about 8,000' elevation. Fire towers are usually located at the highest point on said mountain. Signal Peak lookout is so positioned. There is a "blind spot" to the east of the tower where fpt (fire patrol) personnel have to go in case of lightning strikes in the area. An ancient and very long ladder led to the top of a granite outcrop where the fpt could see down in to the tower's blind spot.

On one occasion, an fpt (old term for fire patrolman in a truck) was doing this job when lightning strike the outcrop upon which he was standing. Knocked him flat off the rock and several feet lower. He was out for some time and came to without help. In the days before cellphones, he could have stayed in that condition for a long time.

The mountain flows steeply from 8,000' down towards the old mining town of Globe, Arizona, with an elevation of about 3000'. Scattered through the slopes are old mining stopes, tunnels and mining equipment, some pieces many tons in weight. One was stamped with made in Belleville, Illinois on it. Ironic, since I graduated from Belleville High School in 1956. Some 1930ish cabins were also in fairly good shape. And a single, solitary Douglas Fir tree stood in isolation adjacent to one of the 4WD roads. Visited only by curious scientists from the universities. A reminder of earlier times.

The old stagecoach road from Winkelman and Kearney ran from Dripping Springs Wash on the south and over Pioneer Pass into Globe. It was on this road that the Apache Kid and two of his companions shot two deputies transferring him and other Apaches to prison. In the sand washes of Dripping Springs, many survived the depression by panning for gold. Historic ranches are fixtures up and down the wash. I remember the Tatums, Browns, Harry Smith, Ed Bearup and Anderson ranches among others.

The mining town of Troy is located just above this drainage. It was active from 1901 to about 1910. About 200 people lived there. It was the only mining town that banned alcohol. It had a band, assay office, boardinghouse, hospital and reading room. And on the same road, the only stagecoach held up by a female and accomplice occurred. Pearl Hart, famed bandit with a sidekick using the alias "Joe Boot" both served time in the Yuma Prison for their deed. Pearl was released after claiming to be pregnant by one the guards.

Of course the historic Apaches were in the mountains, gathering and camping for weeks in the area where large stands of mescal grew. Heated in large rock lined pits, agave roots were a staple for the Apache. These features are ubiquitous in the high desert country. Often found in association with prehistoric ruins. But there is no firm evidence that the Athapascans arrived in the southwest much before the arrival of the Spaniards (1540).

The Ferndell Guard cabin was semi-used by the forest service. A fire prevention technician lived there during the fire season. A much older log cabin with a dilapidated large rock and concrete fireplace.

Close to the fire tower were the usual concrete block enclosed buildings holding the very important communications equipment for various agencies. When lightning hit the Signal Peak Lookout cabin in 1966, all phone gear was melted internally. The phones in the cabin and the tower were useless. The main unit in the concrete blockhouse was no longer functioning, and the crank operated battery phone outside the front door of the cabin had blown off the wall. I found it early the following morning about 50 feet away, across the dirt road and hidden in some large brush. The actual strike hit a phone line about two miles from the tower and merely followed the line into the cabin. I could reach no one except by driving the 17 miles down the winding road to Globe. Came the dawn, everything was replaced as the tower was non- functioning.

Signal Peak Lookout (1966)

My dawn to dusk shift was over and I descended the seventy foot tall fire tower into the small two room cabin. Since I was living at about 8,000', Tucson was many miles to the south and the Mogollon Rim many miles to the north. If I wanted the Tucson television channels, I simply bent the tv antenna south, if I wanted Phoenix - west. Good reception.

One of the drawbacks was periodic lightning storms. You're closer to the source. During one storm, which sounded far away, I was brought up short by an explosion that literally lifted me out of my chair. Inside the cabin the television shut down immediately - simultaneously all of my light bulbs exploded inside the cabin. It was pitch black - no flashlight within reach, the shattered light bulbs hit me with small fragments of glass, but no injuries. I stood up, fell over my cat, found the flashlight and surveyed the damage. No power at all. Lord, it was dark.

I was supplied with three types of radios. The first and main unit was inside a huge concrete block station below the tower. The lightning inside the lines had tracked it inside the concrete and literally fused the insides. Radio number two was the telephone line which the lightning had just fried along two miles of line. Number three was a small, hand cranked, battery powered outfit like your grandpa used on the farm. It was dead. All three were useless. I did the noble thing and went to sleep.

In the morning, there was still no power. The batteries in the hand cranked unit were attached to the outside wall by the door the incoming power surge had blown the batteries one hundred feet across the road and I found their fragments in a small bush. I had forgotten my small walkie-talkie. Replacing the dead batteries, I climbed the tower and called the Globe District (nothing). Called Phoenix, they could hear me. I reported the damage and felt better.

Before the main unit was installed, Rick Bell, who was lookout for years on Signal Peak told me of sitting in the same room while a bolt of lightning-thick as your thumb-shot straight across the twenty foot width between the two door keyholes. If you happened to be in the middle, you were instantly a statistic. This was when the old power unit was located inside the cabin.

During lightning storms, when I was in the tower, I sat on a wooden chair mounted on glass balls for insulation. Remembering not too touch any part of the metal framework or the alidade. During some storms, small blue balls of static electricity would roll around the wooden floor at my feet - the hair on my arms would stand straight up.

In 1966, I was struck three times -some four feet above my head while in the tower. Forget the old adage about lightning never striking twice in the same place. I have seen heavy lightning strikes hitting the same area many times.

On other occasions, heavy showers of large hailstones would drive in at an angle, hit the metal roof and glass windows and I never understood why the glass didn't break, the sound was beyond description. Imagine your head has a metal bucket over it which fits tightly around your ears, then imagine someone pouring hundreds of small one inch steel ball bearings on top of the bucket from about one hundred feet above, that would approximate the sound. And adding to the fun, lightning is zapping in, at and around your tower with great rapidity.

On rare occasions, you get trapped in the tower. Sometimes the severe storms just get stuck in the same place. They don't move. And the very heavy rains and copious lightning continue and don't move on. So, I haul out the sleeping bag. Sleep on the wooden floor in a tower exactly 6' square. Until the lightning strikes become fewer and wait for the storm to move on.

Many "hot" strikes occur. Those are the strikes you see that seem to be much thicker and hang in the air for a long time. Sometimes they strike in front of you. I have seen entire trees explode in front of the tower with a tremendous, loud, crash, and at other times they strike fairly close and there's no sound at all.

On one lazy, windless, hot day like so many, I was glassing the horizon with my 7 x 50 binoculars. I kept hearing a small motor, like a small motor scooter in the far distance. Sounds carry for miles at this elevation. I decided to look through the binoculars closer to the tower and finally straight down to see my cat absolutely frozen in position with a black "timber" rattler some feet in front of her, the small motor sound being created by the rapid shaking of several rattles. The snake was coiled and the cat at least had the sense to stay out of striking distance. Couldn't have a hostile snake within the cabin area-so I called the Ferndell Guard for help. Ferndell was another station with a small, rustic cabin located inside a grove of aspens. The guard drove up to Signal Peak Lookout, and I went down tower in an attempt to locate the critter. The buzzing noise was now much louder. Said timber rattler had crawled into heavy brush which was a smart move- on the snakes' part. Not wishing to pursue the matter and believing discretion was the greater part of thinking, much less valor, we gave up the great snake hunt. The guard went back to defending the woods and by now that snake's descendants are scattered all over the mountain top.

Signal Peak Fire Lookout Cabin
Summer of 1966

TOP: PROPER LOADING
BOTTOM: THREE STOOGES LOADING

MT. ORD L.O.

1970's

Replaced by a much larger and civil live-in cabin. The cab of this tower lies in back of the Rim County Museum building in Payson.

Stan Brown and I were visiting said place and I saw her on her side.

What memories.

Twelve Hours In A 6 Foot Square
Metal Box 70 Feet Off The Ground

Seated on a stool supported by the glass telephone insulators purveying the universe, the Mogollon Rim to the north, Mount Graham to the east, Mount Lemmon to the south and the Camelbacks to the west eternity here, is timeless.

Everything I despise is absent. Time clocks, ten minute breaks, traffic, long lines for anything, anal retentive bosses, kids who use cellphones and text while driving threatening my existence and HDTV.

Somewhere in time we all felt the mystery of life and the darkness of the soil flowed through the fingers and in the genetic makeup of the species we were all destined to be farmers, people whose clock was the seasons of the year and the rising and setting of TLALOC.

Timelessness is here and one's primordial feelings return. Watching the hawks and eagles and even buzzards soar by, the giant shadows slowly shifting creating paintings with each passing moment, the sun slowly sinks through a myriad of delecate shades and when you can no longer see, you descend the tower. Down the wooden stairs with thin metal handrails, feeling the slight sway of the living tower, and the presence of the ubiquitous wind, snap on the big brass government lock at the towers base. Into the cabin, prime the handpump which brings up the ice cold snow derived water collected below the cabin in a giant concrete, charcoal filtered cistern. Start the fire in the woodstove, cook a slow meal, night approaching without the intrusion of city sounds. But hundreds of slowly discernible sounds from microbeings sharing the space with you, begin to increase.

Meal finished, walk outside and scan the glimmering bejeweled city, take a deep clear breath, actually "be" for a time. Back inside, reading, "Only Cowgirls get the Blues", or Ed's, "Monkey Wrench Gang" - now there's a book for all federal employees, or, "God is Red" by Vine Deloria. The head begins to drop, turn off the lights, into the three foot high -off the floor -brass framed bed, each leg grounded by glass insulators (in case the lightning gets inside the cabin) - slip into sleep. A last drowsy glimpse through the eye level window shows the totally silent but shimmering city below. A giant aspen sings ancient melodies to me as the nightwind bends the remaining fragments of light. Wake in the early, silent, cool dawn, greatful for all of life.

The bomber comin' right at us, noisy as hell, old style four engine job. You can tell the wind is nominal (which is really good), and the chaparral brush type terrain will not present a major problem (unless of course the wind suddenly increases, or the bomber misses his target, or the firegods are in a surly mood).

Newbies like to get directly under the slurry/bentonite (gorilla snot) and get their nice new fire shirts covered with the red goo….Proves you been at the head of the fire. Of course, if the more than a ton of red liquid doesn't break up enroute to the ground, it could break your spine like a chicken leg, after all, it does make some really deep holes in the ground at times.

Always loved to see the bombers come in and cool things down. Fires are not suppressed with their action, but it sure slows down the rate of spread and makes it possible to subdue the monster.

Fire Suppression Crew - Sunflower Work Center - Mesa Ranger District

Early 1970's.

Back row: Tim Connelly, retired and living in the Payson, Arizona area.
Paul Wakefield, retired from the Coconino, N. F., Flagstaff.
Scotty Anderson, from Michigan - no information
Russ Copp, forest FMD (Coconino,) retired August, 2012.

Middle row: Bob Pena, retired fire staff, Lincoln N. F., New Mexico.
Dave Jones, retired, living in Lakeview, Oregon.
Jon Selby, law officer on the Tonto National Forest.
Unknown gentleman.
Tom Tillman, retired, living in Colorado Springs.
Unknown gentleman. (Bob May?).

Front row: Unknown, from Oklahoma City, heard he was deceased.
Carl Graves (spitter
Jon(John) Deal(Diel?)-retired from Ariz. Game and Fish.

I don't recall his name, but he received a warm welcome from the Sunflower Work Center fire crew members. Seems he drifted down on a TDY from the Modoc. He died a young man while driving a motorcycle too fast - sometime after he returned to his Modoc Fire Crew.

After several days of double shifts on the fire line, the Sunflower firemen are in need of cold refreshment.

L. to R. Bob Pena, retired fire control on the Lincoln N. F., Scotty from Michigan, Phil Smith (author), Bob Kuhn, now retired – formerly roamed the halls of the FS in Boise, Idaho. Last was a TDY firemen from the Modoc N.F.

Indian Wash Fire

INDIAN WASH FIRE WAS RAGING and Bob and I had slowly walked down a road backfiring all night. Carrying two foot long fusees, or kerosene drip torches, we worked one side of the road. The idea is to have fire meet fire and extinguish itself. In the night time the winds become calm and in the high country, they also reverse direction.

Bob and I were resting, after consuming a typical meal (Kentucky Fried Chicken)- which arrived five hours late, sleepily delivered from the back of a gen-u-wine forest service truck. Eating the ice cold chicken we didn't complain.

My fire friends closer to the border would have received food from "Kontyky Pollo Frito), with restaurants in Nogales, Juarez and Mexicali.

We then checked the other side of the road to be sure no sparks had blown over and started anything when a literal small round ball of fire on small feet scurried across the highway - running from the hot to the cold side. Bob and I, being separated by about thirty feet both ran like hell and met at the small conflagration simultaneously.

On a dead run, we met and grabbed this small ball of fire at the same time. Turning on our hardhat lights we found a small, singed small baby rabbit whose fur tips had been ignited. We patted him or her down, checked him or her out and he or she seemed fine. Putting the bunny down, the animal blinked a bit, than roared off into the brush - this time without any flames. Had the small ball of fire hit the brush behind us, we would have had two fires, one we were making on our left and the new one on our right. No one in the fire business would have believed us the next day that a ball of fire ran across the road and started a new blaze. I hope the bunny sat up nights when he/she aged a bit. What a story it would tell inside some deep rabbit hutch on a clear, cold winters' night.

Friend Walt Sniegowski, former Little Tujunga hotshot crew boss, tells me of one named fire, the "Polecat' escaping due to a brush rabbit and/or woodrats running through a road culvert and starting a significant conflagration in San Gabriel Cyn, the fire consumed an additional 15,000 acres due to these critters- not caused by a polecat, the name was chose because the origin of the fire was in "Polecat Canyon" He continued, "The most unusual animal/fire story I heard of during my career involved a packrat or woodrat. Probably of the genus neotoma. These rodents are notorious for taking shiny, metal or manmade objects back to their nests

One of these critters apparently took a number of wooden-strike-on-the-box matches back to his/her adobe hacienta. While enjoying a "chew' after work, apparently ignition occurred along with self-immolation. F. S. super sleuths (aka arson investigators)found the perpetrators" remains and easily identifiable evidence at the fires'point of origin.

J.R.

Eating lunch in the small cafe on Saguaro Lake, the last lake in the Salt River chain which begins at Roosevelt Lake miles away, we were enjoying a true moment of rare relaxation. J. R.Parmier retired after years of service in the Canyon Lake area.

Weekend Patrol

Fred (a summer employee and principal of a Phoenix elementary school) and I rounded the corner into the campground. We had problems on our hands, one of the county sheriff's posse had his .44 magnum in his hand, braced on top of his car, shaking so severely I thought the gun would discharge. His very lethal weapon was pointed at the recently nude now semi-clothed bearded youngster who had decided further up river to float down the river minus clothing and exit in front of the spot on the beach where sat the officer's wife. Immediate cultural problems arose.

The officer after hearing some poorly chosen words from the young man just about split his head with the gun barrel. I walked over to the volunteer and said, "This man is not armed, could you holster that thing". No response. Walked over to a real officer on horseback, repeated my idea. The weapon was immediately holstered.

Scores of young tubers hurling epithets at the officers, an out of control scene. Walked over to the thoroughly bleeding bearded one and bandaged his head, playing devil's advocate to the crowd - seemed to defuse the situation. Problem was that real officers had tried to share power with people whose mentor was a mix of John Wayne and Barney Fife. Great idea, usually works, this time it didn't. Never met a better disciplined bunch than those with the Maricopa County Sheriff's Department, but that day almost ended in tragedy on both sides,

Killer

Killer was a mule with an attention deficit problem. Like most of the useless equine flesh Smokey seems to buy. Actually, most of the critters were not really purchased, they simply are former "pieces" of transportation – now found to be expendable by the United States Army, by means of a simple transfer of property. These animals were routinely used for trail construction, telephone line installation, packing in groceries for isolated fire lookouts, packing out injured hikers and hunters and carrying barbed wire for fence construction.

We didn't care that it took three big guys to load him(two on both sides shoving, assisted by foul words.) one on front with the nosebar- you had to get his attention first). Killer would carry a large load until just shy of the top of a hill - see the hill - and take off like a herd of wild bees was upon him. The load would scatter over several hills. And we got to repeat the loading process - while filling the air with several culturally chosen words. Once, she tripped over the leadline and came crashing to the ground, her head bent backwards, I knew she was dead.

I yanked the pack off- Killer jumped straight up like a spring colt, and pranced around having won again.Last I heard, old Killer had broken the fence at Blue Point Station on the Verde River and eloped with some of the Gila Indian wild horses. Never had to pack her again, but she did have class. In the late 1960's, on the Mt. Baldy district of the Angeles N. F., one mule named "Francis" was declared to be excess property. Francis was 41 at the time, and according to F. S. records was born at the Missoula, Montana remount facility, was transferrred to the U. S. Army and then finally back to its original owner-the United States Forest Service. Francis was sold for twenty five bucks(the cost of a medical exam) to a gentleman farmer(who had the minimum required piece of property, which was no less than three acres in Chino, California. We think Francis now resides in the meadows God reserves for mules who have crossed over.

Canyon Lake Journal

(Unedited)

Ski races continue, found a three foot long dead rattler in camp, closed Acacia and Palo Verde Campgrounds - both full.

/16 - trash day,took five loads out of canyon lake, gentle light rain fell over the Tortilla Mountains.

/17 - Saguaro Lake with Jr and Stan-hauling trash, a kid's dog scratched his eyelid, treated him at Saguaro Lake Cafe, Stan backed both his truck and trailer into the lake - pulled out via a pumper truck. Saw two over-turned portable toilets heading for the furthest shore- pushed gently by the waves. Heard that five people drowned in the lakes the past weekend.

/18 - Jr and I fed Josephine and her puppies at Blue Point, four horses and Killer, the mule were still running loose on the reservation, more trash hauling, cut some palo verde trees on entry one blocking the boat entry on Saguaro Lake. An empty, really large portapotty fell out of Stan's truck enroute to Bluepoint - luckily, no one was following him at a close distance.

-21 - youth on motorcyle minus helmet - warned him, find a helmet, or get off the road. People in and out of off limits boat dock (used to be Ockletrees). Someone asked me where the extinct volcano was? And where was Spoke Mountain??? Told civilians to get off the one way bridge at Boulder.

-22 - East Sunday. Lunch with deputies. 16 year old drowned in Lake Pleasant yesterday. 7th one this season. Patrolled all morning. Lady lost her big black dog at Fish Creek last night. Minor car wreck at 1:30 pm. Two girls playing tag with their cars.

-23-Trash pickup took all day. JR put the left rear wheel of the trash trailer into the garbage pit, but we saved him. 3pm truck died--battery dead.

Next Weekend

- get more garbage cans to block illegally parked cars under Acacia Ramadas.
- Asked five people to move their illegally placed cars under Acacia.
- Someone destroyed the official government lock at Boulder, i rewired it, no spare. Man didn't want to move his boat trailer, parked illegally, said that someone might steal it.
- Man griping: Kids on the hwy. above throwing rocks down on us. Sought them, but could not find.
- Man: you need separate tickets for a car and its trailer?
- Man: Can i put in two dollars and stay two days?
- Smith: "Chain your dog sir,(his loose dog almost bit two boys)./
* Closed the Ochiltree dock again, ran twenty people out - three boats docked there.

Ticket box empty at Palo Verde, many cars without permits.

I asked four "Hells Angels" surly types to move their hogs, they did so with great reluctance.

Deputies numbers, Carlson(190), Tyree(192) and Edwards (196).

April 28th to May 2nd, 1973)

A young man shot himself in the balls while loading a black powder, cap and ball pistol, he immediately blamed the forest service. The chopper was called, but took too long to arrive, so his friends grabbed his frame, hurled him into the back of a small pickup truck and lit a shuck for the big city and some competent medical people. Never heard how he fared.

Boy: where can i get fish bait?

Two men: a black "65" chevy roared through camp, hit an 8" x 8" post, spotted them roaring into Tortilla Campground. Roared toward us. obviously under some influence. They had a blowout at this time and were parked when we got there. They hailed us. One man had a huge head wound, another man (with tire tool in hand) said the first person had struck him first. Both claimed it was "a accident' and said, "jail us', but first, phone our wives. Since all the deputies were involved with more serious business, we let'em go.

Smith: phone Pierre.

Smith: man drowned at Blue Point last night (copied it over the MCSO radio).

Smith: deputies door kicked in again last night.

Smith: Someone pulled a gun and was firing at the skiers on the lake. Deputies confiscated two guns and lots of fire water.

Smith: All lanes at Biesmeier point were closed off to all boats trying to get in the lake, some of the rich, obnoxious boat owners said they preferred it that way. "Make an opening please, as in now".

Smith: Another drowning last night.

Smith: Ran another seven cars out of Acacia Campground.

Man: Do you carry the golden passes?

** Fred was collecting tickets at the entrance to Saguaro Lake. this elderly woman drove through after getting a ticket from Fred and drove straight as piss into a parked car. a bit shaken, she exited her car, walked back to Fred raising hell and saying that he had caused the wreck. Wonders never cease.

Smith: car accident by entrance to first bridge at Canyon Lake/wheel fell off the vehicle.

Smith: Reminder: for the next five days, two thousand sheep will be passing through the Blue Point Recreation area(over the bridge) heading north on their annual trek - crossing Sycamore Creek and then walking north to Reno Pass ending up in the high rim country. Basque shepherds with their bell carrying mules do the work. Talked to one of them. My knowledge of the Basque dialect is equal to my knowledge in Algebra I

Smith: Body not up from last week's drowning.

Smith: Re-drove the NO PARKING signs which someone knocked over.

Smith: Someone stole $70.00 out of the money holders at the parks' entrance (probably the old bent wire trick).

May 6th - Saturday (8 am).Clay wants JR to see if the dogs are loose with the sheep at Blue Point. Fred is NOT to be near the ticket box at Saguaro Lake as the father of the boy who shot himself in the groin may be looking for him.

Smith: some tour jeep with humans, one dog and one red gasoline jerry can cruising thru the open desert without benefit of a road.

Smith to man: "lease your dog, sir"

Smith: camping teacher from ASU grouped in Boulder, drying out all their camping gear--but the cars had no way to turn around.

Smith: About 30 kids and 3 adults crossing Boulder Bridge - all ran - scattered to the winds.

Smith: Fishcreek got 1.2" of rain, Tortilla Cafe, a half inch. Water over the road again at Tortilla. Rain clouds covered most of the lake.

Smith to man: "Leash both your dawgs, please".

Man drove up and said there were dawgs running around, I timed this one beautifully. "Sir, the problem is resolved).

* Someone broke the lock at Laguna Camp again.

* resupplied the contract man

* driving time from Tortilla entrance to 03 yard is fifty minutes.

Sunday, May 5, 1973

Fire patrol duty, no one at Cottonwood Ranch, met Glen Randall working on the Circle Bar outfit.

* Lots of quail and doves this year.
* four money vaults broken into on Canyon Lake today.
* Lock at Laguna is broken again.
* Alternator fell off the truck again (third time).
  (A Datsun 240Z was totaled by a Buick by Water Users Camp).
* Talked to shepherds with 1870 head of sheep, they were stopped at Sycamore Creek, as it was still two feet above normal due to the rains.

May 6th - Saturday (8 am). Clay wants JR to see if the dogs are loose with the sheep at Blue Point. Fred is NOT to be near the ticket box at Saguaro Lake as the father of the boy who blew his own balls off may be looking for him.

Smith: some tour jeep with humans, one dog and one red gasoline jerry can cruising thru the open desert without benefit of a road.

Smith to man: "lease your dog, sir"

Smith: camping teacher from ASU grouped in Boulder, drying out all their camping gear--but the cars had no way to turn around.

Smith: About 30 kids and 3 adults crossing boulder bridge - all ran

Smith: Fishcreek got 1.2" of rain, Tortilla Cafe, a half inch. Water over the road again at Tortilla. Rain clouds covered most of the lake.

Smith to man: "Leash both your dawgs, please".

Man drove up and said there were dawgs running around, I timed this one beautifully. "Sir, the problem is resolved).* - Someone broke the lock at Laguna Camp again.* - resupplied the contract man

Monday
* ran out of forest keys, mine's thin, my partners lock is broken, had to cut hole in roof of the shed to get in.

Saturday May 12th
* - collected money at Acacia Campground
* - fined four in Biesmeier without tickets.
* - towed stuck truck and boat trailer out of Biesmeier.
* - have you purchased sign at Biesmeier was destroyed.
* no parking sign knocked out of the ground at Acacia.
* man with boat trailer taking up 3 spaces at Acacia.
* - all tickets gone at Palo Verde box, and the lock was open, I relocked it*ant beds by PV need some Greenlite.
* - At Laguna, the money was stolen from the vaults, one man gave me three dollars which was on the ground. Replaced Laguna lock with a broken lock (it will be anyway - probably overnight).
* - Wienel leaving Tortilla Cafe putting up signs at Reavis Ranch.
* - Deputies demonstrated the resuscitator in the deputy station

* - lake patrol with deputies - middle buoy under Boulder Bridge shot to hell. Beer Can Point very clean.
* Three golden eagle pass inquiries.

May 13th, Sunday

* replaced Laguna lock.
* replaced lock on Tortilla shed.
* "There's a eucalyptus tree on sire leaning on some hot lines!
* Closed off bottom ramp at Tortilla for the season.
* five cars illegally parks at Acacia C. G.
* No tickets in the box at Acacia.
* 12:30 pm - joined posse' and fish and game gentlemen in joyful walk through the diving rock, was quiet.
* four vehicle accidents between Saguaro and Granite Reef dam today.

May 14th, Monday

* no fire on north side of four peaks, those are water dogs.
* Blackwell broke my new forest service key trying to unlock a pissoire.

May 15th, Tuesday

* hauled trash
* fires all over - Payson, Pleasant Valley, even Cave Creek.

May 16th, Wednesday

* worked in lower Salt and Butcher Jones C. G's until 11am.
* Sunflower store at noon, with Les until 1pm, Cottonwood road at 1:35 pm.
* Smoke report on Indian land by Verde bridge on the Beeline. Five people arrived at the same time. Grunewald, JR, Weinel Clay and I let her go.
* - replaced tickets at Del Norte.
* - engineers destroyed the gate at Blue Point entrance.
* - Jrs' truck died enroute home, everybody gone, nobody copied.

May 19th

* goofed a radio transmission.
* some group is parachuting into Canyon Lake today.
* Fire on the Coconino and in Payson today.
* 3:15 pm - girl drowned in Salt River, two men found her face down, underwater over five minutes, chopper arrived quickly- set down on the gravel-mid-stream. Deputies used an entire oxygen bottle, but she had the fish-eyed look. She was a mom from New York, a youngster was going under

in front of her - she dove in, saved the kid - and died. Never even made the evening news.

May 20ᵗʰ

* - man on motorcycle seeking shade, it is to laugh.
* - another Golden Eagle inquiry.
* 9:30 am - someone's firing weapons at Water Users Camp.
* Blackwell said Laguna C. G's money was stolen again. Seems a man is out on bond from Payson robberies. Word came through: Be on the lookout for an old red Ford truck with a rebel flag flying from it, and oh yes, he has a three legged dog - should NOT be difficult. He disappeared the night of our robbery and was caught up in the rim country.
* Grunewald's hood won't open except when he's driving down the road.
* - returned to Stewart Mtn. Dam, Deputy required aid, people wouldn't leave. when I arrived there was no one there.
* - to Deputies station(Canyon Lake), eleven people arrested - fighting and drugs, I treated one cut leg, helped fish/ game guys unload the fighters.
* - loud screaming in the interrogation room.
* - hysterical girl sounds through the walls - then angelic quiet.

May 21ˢᵗ -

My contract ends in December. Sunflower may be phased out. took law enforcement class at Mesa Dist. Office/ Walters instructor.
* fire patrol on lower Salt River.
* - two males in purple Datsun driving one hundred mph chasing young white male in Bel Air who stole everything from their car.

May 26ᵗʰ -

* - Fire Patrol on Bush Hwy and Salt River
* - Man: Where's the Salt and Verde Rivers meet? Phon D. Sutton
* - Tonto Basin R.D. had a small grass fire.
* ran off the road when the truck died pulling a steep grade, set fuses. Pete Weinel and I on initial attack over Usery Road. Beekeepers set fire to brush when smoking bees - about 1/4 acre fire.

May 28ᵗʰ –

* report
* - leave for Salt Rio Fire Patrol
* - coffee at Sambos in Apache Jct.
* - load 75 gallon pumper trailer and back pack pumper/Blue Point.
* - closed Saguaro C. G. - huge line of vehicles.

*   - took two aspirins, boy fell off motorcycle, skinned badly
*   fires everywhere today, San Carlos, Wheatfields, Six Shooter in Globe, Payson, Bartlett Dam and the open area north of Phoenix. (bombers used).

May 29th

Met new man on TDY from the Angeles National Forest (Walt Sniegowski)
Additional
*   - Gary Butterfield was shot in the neck at Blue Pt. Bridge by CBers.
*   - Fleischmans' son rescued south of the bridge, almost drowned, chopper couldn't be used as he was too violent, ambulance was very late.
*   - another drowning
*   - continued fire patrol
*   - rained from 11am to 3pm/off and on at Tortilla Mountain / Fish Creek/ all of Canyon Lake/ much lightning.
*   - they fired the lifeguard due to lack of performance, said he paid his motel bills, but he didn't... also broke headlights and stole money from campground people. I removed the lifebuoy sign, canteen, resuscitator and tools from his room with his permission.
*   - Diving Rock to be closed next weekend/ 72 felony arrests in one month
*   - filled ticket boxes on Thursday, got the money, saw Walt Sniegowski, got deodorant cakes for sub-contractor, he broke my vehicle mirror driving down wrong way entrance to Laguna.
*   - got wheel barrow and sign for Stanley Eugene at Blue Point, put up road closed sign, rained four hours, patrolled to massacre grounds and collected the wilderness cards.

June 19, 1973

*   - arrived canyon lake, collected money from Acacia C. G.
*   - replaced tickets, collected loose cans in area, gave contract man to bug sprays, he needs more plastic liners, Beismeier keeps removing the deodorant cakes.
*   - checked area, man with boat trailer needed ticket, he got one
*   - no tents allowed in the area
*   - man: where's the drinking water?
*   - womans bathroom sign stolen at Biesmeier, screen also knocked out.
*   - Two men: is the fishing good?
*   - wrote personal note on car for man with Golden Eagle passport.
*   - Wrote citation on huge Cadillac with boat trailer.
*   - two girls: do we need a permit? Woddya want to do?
*   - asked for permit from Illinois car.
*   - three garbage can lids blew off at Laguna C. G.

*   - man showed his ticket(guilty conscience).
*   - 2 pm - prepared 42 lids for painting, swept out Tortilla shed, got six full trash bags from east side of Boulder picnic area.
*   - break at Tortilla cafe, really hot today.
*   - 3:40 pm - emptied deposit boxes, two bikes and six cars under the ramadas at Acacia - ran'em out. OK Corral man in there with chopper friends - More kids on the bridge, asked to leave. Biesmeier okay, all have tickets, dogs running all over hell minus leashes.
*   - returned to 03 (Mesa District Office).

Smith: Re-drove the NO PARKING signs which someone knocked over.

Smith: Someone stole $70.00 out of the money holders at the parks entrance (probably the old bent wire trick).

May 6th - Saturday (8 am). Clay wants JR to see if the dogs are loose with the sheep at Blue Point. Fred is NOT to be near the ticket box at Saguaro Lake as the father of the boy who blew his own balls off may be looking for him.

Smith: some tour jeep with humans, one dog and one red gasoline jerry can cruising thru the open desert without benefit of a road.

Smith to man: "lease your dog, sir"

Smith: camping teacher from ASU grouped in Boulder, drying out all their camping gear--but the cars had no way to turn around.

Smith: About 30 kids and 3 adults crossing boulder bridge - all ran

Smith: Fishcreek got 1.2" of rain, Tortilla Cafe, a half inch. Water over the road again at Tortilla. Rain clouds covered most of the lake.

Smith to man: "Leash both your dawgs, please".

Man drove up and said there were dawgs running around, I timed this one beautifully. "Sir, the problem is resolved).* - Someone broke the lock at Laguna Camp again.* - resupplied the contract man

Billie

It was early summer in 1966. I had finished my second year of teaching choral music at Globe High School in Globe, Arizona. I applied on the Tonto National Forest in Globe for a summer position.

Bill Buck was DFR(District Forest Ranger) then. Ex marine with a marine haircut, tall, cowboy legs - no nonsense-hell of a man. I won the job of fire tower lookout located on the Pinal Mountains, the tower was named Signal Peak due to its being used in the post Civil War period to signal other military units as to the whereabouts of Geronimo, and eventually helped subdue the warrior.

Elevation was over 7,000'. It was one of the "old" towers. That is the cab in which I worked was about 6' square, not counting the range finder in the middle of the work area. Super thick glass, metal framework, metal roof. Each leg was driven deep into the ground, the bottom part being a giant ball of copper wire.

Grounding wire, without which you would be dead when struck by a million volts striking the point 5' above your head. Which meant I was to work from dawn to dusk in a cab 6' x 6' with a walking area of about 2' wide and 70' off the ground. My tower on top of Signal Peak was on the extreme southern edge of the Tonto National Forest (Globe District). Working tools were simple and few. A pair of government issued Navy binoculars, and two or three different radios with which you phoned in various information at set times during the day. There was a land line anyone could call, a forest net line with which I could contact all districts and head office in Phoenix simultaneously, and an ancient, battery operated WWII phone(crank operated) fastened to the outside wall of the cabin, plus a second landline inside the cabin

There was also lots of time. You were tower bound until the short run like hell down the shaky tower steps to the outhouse time – always afraid you'd hear, "Signal," and nd not able to respond.

Not city time. Fire tower time. To say that time moved forward in a slow circle would be one analogy. To say that when the heat arrived in July and August, time hung suspended like the night (too flowery).

Some lookouts that were as new as I simply suffered withdrawal and other symptoms the entire two weeks of their brief tenure and drove back to Sheboygen in the middle of the night, leaving the tower unmanned (unpersoned?)

When you are looking for smoke of any kind, you must do a 360 degree circle with the binoculars. They are a bit unwieldy but the very slow pivoting covering 90 degrees continues. A lookout is supposed to cover each ¼ circle every fifteen minues, which means you have time for nothing else within the 60 minute period.

For some reason, I had spent a good part of the morning looking north towards the Mogollon Rim(seen in the far far distance) and west towards the Superstitions(looking over the dessicated land destroyed by the copper companies.) I rarely looked to the south as Dripping Springs valley was well below Pinal Mountain and the southern edge of the district. This tower was almost the southeastern landmark for the Tonto. The Dripping Springs area was pure upper Sonoran desert, cholla, palo verde trees, sacahuista, bear grass, prickly pear and other cacti, but not a tree (ponderosa or oak) could be found. So the fire danger was minimal.

Got a phone call from Billy, a summer cabin owner in quiet, low bass tones (she had a deep, husky voice). Conversation went like this: Billy," When ya gonna turn in the fire?

Me! FIRE, WHAT FIRE?

Billy: The one that looks like half the world's been napalmed, down there in Dripping Springs country!

I looked south and about fell off the stool. A raging fire, bout a half mile wide was roaring away from the tower on the south side of the giant wash down in Dripping Springs. I checked the rangefinder and saw that it was a good mile off the forest on state land. Phoned it in anyway.

Friend Tom Dodson (one of the oldtimers) took a crew down the back road through Doak (an old mining cabin and spring which were visible from the tower as of 1966), I should say they built road all night, moving giant boulders and fallen trees, filling in giant washed out chuckholes, and by morning, when they arrived, the fire had consumed itself, as many do - simply ran upslope into heavy rocks and out of fuel. Tom was not in a great mood. The only way back to base was back up the trail they had spent the miserable night creating.

The redtail hawks would fly in like a jet on a bombing run, stop about fifty feet from me - catching the always present ridge wind, and just hover, their bodies still, their trim feathers flapping up and down, adjusting their tail feathers with their head moving back and forth, lookin' for lunch. They would stay in this position, motionless for what seemed like long periods of time.

But the comedians of the air were the buzzards. Those highly maligned funereal characters. I could see in the far distance a slit of black wings coming at me like a straight winged World War II fighter plane. Gathering speed, these clowns would head directly at the tower, like feathered kamikazees, until at the last second of their strafing runs, they would peel off, I would fall off the stool, and they knew they had strafed me into humiliation. I swear I could hear faint vulture wheeezes as they roared by my lookout tower.

Sun slipped out of sight and I descended the tower. Noted since it was August that thousands and thousands of ladybugs covered the brush - it was their breeding season. Used to be that many people in the Phoenix area would drive up, scoop them up in large cardboard boxes, tape them shut, and return to their gardens and farms - letting the critters out. Great little aphid hunters. Was a natural non-chemical way of killing the insects. Found their little round black and orange bodies in my ashtrays, carseats and every crevice of my 1964 Mustang for years. No one seems to know what male bugs are called. He bugs? Heshe bugs?

The wooden, front screened door cabin was fairly old. Heat and some food preparation was done on the wooden stove. Plenty of wood already sawn and piled.

Water was a 100 gallon metal cylinder hauled up for the season. There was an underground reservoir with which to catch snowmelt runoff, but it was empty. The ancient spring metal framed bed was beyond supporting the normal spine, but functioned okay. Each of the four legs was placed into a single glass insulator as one sees on old telephone lines. A simple wooden table, some ancient chairs and single light bulb in the ceiling (for night time) were the other artifacts.

Giant black beetles(Pine Beetles) 3" to 4" long covered the screens. Night sounds fill the quiet air, some sinister. Sleep is long and morning comes early. During the day, visitors are few. An occasional employee might be trail clearing, painting whatever needs it, or just checking the various campgrounds for compliance or simple garbage pickup. Few ever climb the tower. I've known gnarly fire bosses who have such a fear of heights, I had to climb down to the tower base to get their message.

Weeks can go by in the endless, surreal stream of time passing before a fire is spotted. Crews are called out, sometimes bombers and choppers dispatched and we watch its'progress from many miles away. It is not uncommon for a dry lightning storm(lightning hits the ground without any moisture) to start hundreds of fires in one giant sweep of the forest. The greater percentage burn out immediately. Usually, the dry storms are followed by enough rain to extinguish the fires already started.

Boulder Fire

We were shovel scraping a three foot wide line to mineral soil and cutting overhanging branches with chain saws all day. Five of us down in a cactus covered clearing. It was unbelievably hot and out of my eyes corner kept seeing brilliant lightning flashes. Too close for me. The usual late summer afternoon storms were sweeping the land and we were stranded on a large mesa top, without trails, miles from nowhere, as usual.

When the lightning strikes got a half mile away, we abandoned our three foot wide, mile long line just as a drenching downpour arrived.

We ran back thru our just cut line (now full of water), soaked through our boots, cold, miserable. We knew we were prisoners on the hill-no choppers available -couldn't land anyway with the lightning and the rain.

Temperatures ran in the 110's that day(until the rain). Nite came too quickly. Our worry was spending the night at a very high elevation when the temperatures could drop fifty degrees. Pnuemonia was my worry - dripping wet, night coming on fast. Sleeping wet.

Just as the rainfall diminished to a mist, I heard the "whup, whup" of the chopper coming in to load us and get us out of there. There are beautiful remembered sounds in ones' life, the sound of the one engine Hueybird was one.

A chopper becomes a living thing. Almost graceful, banking in a big turn, nose tilted forward, circling, soft wheel contact, and simultaneously, the tremendous hurricane like wind it creates -keep your back to the chopper boys. Hang onto your headgear. Climbed aboard, strapped in, the ship rose slowly at first and then gained altitude quickly, full of noise and vibrations, and the junipers became little dots on a landscape. We flew back to base camp. Showers, basic food and some needed sleep.

Claude

Claude was a close friend and brilliant paleontologist. (He passed on in the very small community of Afton, Oklahoma - living with Hazel – where he managed an elegant little farm.) We were in some dusty wreck driving down the dusty, jawteeth jarring dirt roads north of Globe City, Arizona lookin for geodes. (round rocks sometimes with marvelous amethyst and calcite chrystals inside). Headin' for a spot that is now closed to non-natives somewhere in the vicinity of the great Coolidge Dam which slows the Gila River coming from New Mexico and on the San Carlos Apache Reservation. I had to sit on an ugly, splintered box as there was no room for both of us and the gear. Bouncing up and down - this way and that way in the normal rhythm of driving on an Arizona dirt road. Finally asked Claude what was inside the box that my fanny was smacking up and down and he said, "them's blastin' caps". I tried to sit as lightly as possible and ignored the possibility inherent in his words. We blasted out some geodes later on in the day and found some beauties. Claude would go anywhere to find rock specimens. He told me of banging with a prospector's pick on some ceiling while twenty feet away chunks of the same ceiling were crashing to the ground. Another time he climbed up some cut out areas to a high point in a very dark cave. Placing his foot carefully in the cut out holes, he climbed up to realize that there was no way he could see to get down. His partner beneath him had to grab his shoes, one at a time on descending, and place them in each hole. Not swift. That's why people without a death wish never enter a cave alone or without having at least three separate sources of light, matches well wrapped, a battery driven light on a very good hard hat and candles to be lit in emergencies.

On another occasion, his light went out while he was exploring a cut in the rock - miner's call this feature a stope. In the blackest of black caves, he slowly became aware of another being also breathing slowly - close to him - sharing the blackness. You have never experienced blackness until you have been down about four hundred feet and snuffed out your light. I have. Helplessness would describe Claude's feeling at that immediate point in time. His thoughts jumped fast forward to cougars, bears and God knew what. Suppressing a great urge to scream, a clatter of hooves and a giant "HEE HAW" about stopped his heart. The rocky mountain canary (as the miners called them) knew where it was. How a mule could see in the dark was beyond imagination. How in hades this creature wound up way below ground level was not clear. The mule was as surprised to sense Claude in his world as was the opposite case. Claude used to laugh about that one, more in embarrassment than in the recalling.

The first time Claude and I descended down to the 400 foot level, it was in Picher, Kansas. I climbed into a one man - giant banana shaped piece of iron, with enough space for one man - threw my leg over, got in and someone said, "hang on" and down I went- the lower I went, the wetter I got. and there was nothing to hang onto.After what seemed like part of a day, suddenly, the cable lowering me caught slack and gently placed me at the bottom of the great unknown.

I stepped out of the potential tomb and it was like being under a gentle waterfall and very dark. but as my eyes adjusted to the minimal light, I could see a single strand of light bulbs above the floor following the tunnel,

I was looking for minerals called sphalerite (blackjack and ruby) and cubes of galena and large calcite crystals.

Claude and the mine owner came down next (even though there was really room for only one, but Claude was skinny) and exited the metal tomb. We three walked a bit to a jeep which I hadn't noticed. Chink fired up the vehicle and we drove about a 1/4 mile underground hitting large holes and throwing us into the air while I noticed that at times the water was half-way up the tires. He finally stopped us in an area, put both of us in certain positions and said I'll be back. He roared off. Claude and I were separated by at least a hundred feet. I immediately found a bunch of large, square lead (galena) cubes which I began removing very carefully.

After about fifteen minutes, I heard a diminutive "whump" and the driver pulled close to where we were. He said, "do you wanna see this?" so we jumped into the jeep, drove around, got to a large white coated ceilinged area and he grabbed a bunch of dynamite with primer fuse and blasting caps attached. I was looking for crystals and found a good area and climbed up, bent my body around a corner and looked into a pocket where about eight sticks of dynamite were about three feet from my face with the fuse having burned to within three inches from the powder. I did a quick breath, slowly backed up, got into the jeep and wanted out of there.

We drove back to the original area and were eating lunch when Chink said, "see that giant pile of jumbled rock over there?, well about a week ago, me and a James were blasting in another area and decided to eat lunch and as we were feedin" our faces and this whole damn pile of rock came clumpin' down right over there during lunch. You know we have many superstitions down here and one of 'em is we never eat lunch in the same place twice and we ain't eatin" lunch today where we was on that day."

Later in the afternoon, Chink was pushing some giant boulders from a large pile of rocks and slipped and broke both ankles, Claude and I hauled him in great pain to the metal bucket and yanked twice, the monster ascended quietly and Chet disappeared into the upper world. Claude and I gathered our stuff and went up one at a time after that. I asked Claude who in the world was runnin' this ancient elevator and Claude said, "Ralph was". Turned out Ralph had suffered a major heart attack about one month before and was in his late seventies.

As I was slowly ascending through the dark, very wet, vertical tunnel, my thoughts ran to retired Ralph with a bad heart. What if Ralph had suffered a seizure when we were down below? Arriving at the top, where the sun was beautifully bright and warm, a sense of well-being returned.You can't go down the deep shafts anymore. OSHA has closed all of the mines and the pumps used to keep the mines somewhat dry no longer run.

Crazy Jake

Never met him, but I guess he's still camping, hiding out in the Superstitions. I could see his scattered tents and gear far below me from the plane. Stan Grunewald, a fellow forest service employee and I had just flown north of Weavers' Needle (named after Pauline Weaver, one of the early mountain men), and turned north and there she was, blending in perfectly, the square tent shapes were easy to see.

Paulino Weaver born Powell Weaver, born in Tennessee to a "white" father and a Cherokee mother. He worked for the Hudson's Bay Compnay in Canada and in 1830 traveled with 50 other men on a trapping expedition. This took him to Taos, which became his home base. The people of Taos changed his name from Powell to a more familiar Paulino, later becoming Pauline. In 1831, He traveled through Arizona. Later he was Chief of Scouts in 1862 during the battle of Picacho Butte, and was seriously wounded. He died in 1867 while on duty at Camp Verde, and was buried close to the Verde River. Ed. Note: I excavated Camp Verde-

1865-1871 for a number of years while at the Orme School in Mayer, Arizona.

Jake had been in the hills too long looking for the Lost Dutchman mine of international renown. He would come into the Mesa Office and say, "D'ya want any gold nuggets? got some big as my fist this time! The office types would do their best cheshire cat imitations and Jake would vanish quickly into the hills.

The new federal rules don't allow exploring for minerals as in earlier days. The Crazy Jakes will depart the hills like the Fortyniners and the Indians and the small rancher. One of Jakes co-horts shot somebody to death and covered his body with large stones. The law found out. Chopper trip full of borrowed prisoners, they retrieved the body under the stones, and that's the last I heard.

Fed. Mining rules: you can still explore the federal lands, what you cannot do is stake a new claim. The mining act of 1924 has withdrawn much of public land to mineral entry. Unless the land has a "mineral" of tragic importance to the United States or the "mineral" locator is a big contributor to the political party in office at the time (google: teapot dome, solyndra, pot growing, etc;;)

Scores of neat stories about this area, of the former opera singer with a retinue of guards living in a shack back in some canyon for years, of the two hunters sitting by campire being attacked by a huge black man in loincloth and spear, of skulls found with bullet holes, of people going in and never coming out. The number of unexplained deaths is high. Most from freezing or dehydrating or falling while rock climbing or shooting themselves in deer season.

I saw two of the last prospectors coming out with loaded burros. Two short, powerful, unnamed men, no longer young, thinning hair,their eyes spoke of severe isolation and mistrust - written on their faces, was caution.

Looking at black and white 1880 photos, I see the same eyes and toughness of the original settlers. Reflections of a rugged existence, not quite tamed by urbanity. In these two old whiskered men, I saw the last of the breed.

Diving Rock - Canyon Lake]

It was about 1972 and workin' at Canyon Lake, a part of the great Salt River Project northeast of Phoenix, I drove by the geological feature named, "Diving Rock". It's name provided its' function. Just secluded enough and off the road for a bit of privacy.

Younger, long bearded, barely dressed types would cover the rock like seals in the arctic during breeding season (which for Homo Sapiens is year round). It was around the time of the first war we lost and men in any uniform were not objects of deep affection. In addition to forest service people, there were county deputies who patrolled on boats. They patrol in and out of the same nooks and crannies that everyone has sought out for privacy. People go to the lakes to be alone.

They also go to get frisky, drink suds while underage or smoke silly weeds. That's why we have officers to stop the illegal smoking and drinking.

So their little boat cruises in and out of coves searching for these most wanted violators of the law. And smokey finds 'em. Back in the boat to diving rock, as we pass it, a rock big enough to kill an officer of the law rips thru the thin top of the canvas just missing the head of a deputy.

Not funny. Coulda killed him. Regroup: an idea is born, "Clear the Rock": So, a simultaneous amphibious and land assault is carried out. Hit'em from the bridge on cars and from the lake on foot and attack lakeside from the boats. Charge. The fourwheel drive peels out across the bridge, the boat motor screams and speeds on, loudspeakers on board announce, YOU MUST LEAVE THIS AREA IN FIVE MINUTES OR BE ARRESTED.

Moans, belches, digital salutes and hands busy grabbin' possessions, girlfriends, remaining beer and assorted canine associates (four legged kind). Like a sea of quickly moving lemmings to the sea, they head enmasse towards the bridge and their vehicles. Two hundred basically benign potheads punished for one stupid act. People gone. rock clear, except for palo verdes, broomweed, assorted cacti, crushed beer cans, and residual roach clips, scattered mid dog droppings.

The Weekenders

And there they were as we drove into the campground, refugees from the ugly city. Come to get away from it all - all that crowding and traffic and noise. At least in town they are protected from each others madness by fences, here, it's belly to belly and back to back. HOMO SOCIALUS.

Sittin' in their lawnchairs drinkin' suds under a scrawny, emaciated mesquite tree. Out of the city and a lake full of cool water, into which they will never place a foot. Spent $25.00 for gas, $40.00 for food, $87.00 for beer and $1.00 stuffed in the little metal box for the privilege of turning healthy cells to melonomas in the coming decades, three feet from another group doing exactly the same thing. In the big city this closeness would cause homicidal behavior.

Little kids run around naked, young parents leave poop filled diapers all over the campground, one man's dog trots over and pees in the middle of a blanket owned by someone else, someone's on a hill throwing rocks and the escapees in the camp area, and what are we gonna' do about it?

Another man tinkles in broad daylight - oblivious to everything. Some other guy hit his wife in the nose this morning, two waterskiers said someone's firing rifle shots at them in the lake - while they're skiing, another man enters with his young daugher, fishing plug hooked through her ear, he is wild eyed and trembling and has his hand's on the plug.

I say, "Sir, why don't you let me take care of this? The little girl is about eight seconds from losing an ear lobe. I take a pair of wire cutters, snip the back of the hook and simply remove it. People get nuts out here in the sun. Heavy skid marks on the road, vehicle about 30 yards downslope, check it out - blood everywhere, but no humans.

Heavyset sweaty lady chews me out because she's a "Gold Star Mother" - lost a son in a war - and says, "you're charging me a damned dollar to sit in this camp"?- "Yes Ma'am."

Slowly round a curve - spot four people - two couples. Drive about a hundred yards past them, Bill looks at me, I look at Bill. Heavy brakes applied. I do a u-ey. Approach the people - they're minus all clothing. Two marvelously chested ladies sit innocently with their (found out later) husbands. "Morning" - Oh! "Good Morning" Uh, ladies, (we speak very slowly as we gaze) there's a regulation about sitting naked in the sun, even with your husbands, in plain view in a national forest - Jeez, Bill, what code is that?

My fat little fingers are flowing through the Code of Federal Regulations. The ladies responded with a good natured, "Oh, we had no Oydeah" All were from London. And I thought London was strait-laced.

With that settled, we S-L-0-W-L-Y drove away and I looked at Bill and he looked at me and not a word was spoken. Strong exhales. Maintain your professional demeanor The first rule of this kind of confrontation is, "Always look a nudist in the eyes"

Canyon Lake Patrol

Canyon Lake bridge next, three hundred juvenilesexplode from its metal beams and crash into the water far below - ah! the green truck syndrome.

Later in the evening, I heard of a group of girl scouts around a campfire, singing songs with disaster being close at hand. Someone well lit internally but having no external running lights operating in the dead of night hit the shore line, careened through the girls encampment, then through the campfire and stopped. The boat actually had to travel over one hundred feet in the air to get there. The pilot was found in some bushes, semi-conscious. He had been running the boat at full throttle in the dead of darkness without running lights. The girls collected their wits and went home. Never heard what happened to the driver.

In the afternoon, a man hails us to one side of the road and asks us if we change tires? My thoughts were not printable. Continued driving through assorted winding roads, campgrounds, and dusty roads around Canyon Lake. Another normal eight hours out here. Drove back to base in Mesa, Arizona, took about 40 minutes. Parked the semi-reliable old truck and waited until the dawn of another day. When the foibles of mankind would repeat.

Firehose

Ray was jumpin' up and down and yellin' as usual saying, "Get that damn tanker down here!" Bob and I obeyed, we had both just finished a fire year on the Angeles National Forest on a TDY (tour of duty) and we were back on the Tonto National Forest northeast of Phoenix, Arizona.

Now the Angeles has some wonderful stands of ponderosa pine in the high country, and the entire countryside is built on severe slopes, when you get an inferno in that country, it is radically different, and far more dangerous in most cases, than a fire on the Tonto National Thicket (our name, since so much of the two million? acres lacked trees, or even serious brush.)

The first rule of fighting fires is drilled into new smoke eaters. Always have an escape route when entering a fire situation. More recently the ten basic fire fighting orders have been summarized with the acronym (LCES).

l= Lookouts, C=Communicating, E=Escape Routes, S= Safety.

Since we were peons, we moved as ordered and watched the fire round the curve, hook again, and head straight at us- we didn't wait to get the hell out of there, so we kept the truck safe, and watched a thousand feet of one inch hose evaporate in the flames, so we preserved the ugly, semi-green fire truck, and left Ray jumpin" up and down and screaming, which is what Ray did best.

The History Of "Forest Green"

Original F. S. vehicles were excess to the needs of the U. S. army. Some were camo-painted or olive drab, larger construction vehicles were a very dark green color.

F.S. adopted that dark green color until the 1950's, when it adopted a two-toned light green body and a light gray colored roof. The doors were painted or Decal-ed with an I. D. #(I.E. 5-4(engine) or P-5-4(patrol or prevention.)

To reduce cost, the gray roof was changed to the single color light green in use today- which is called by the agency Forest Service Green.

To set the record straight, the color is not Forest Service Green, but is identical to the color Celeste Green. Celeste Green is nothing but the world famous color of Edouardo Bianchi's Classic Bicycles (manufactured in Milano) and ridden to triumphs in the Giro D'Italia and the Tour de France.

Fausto Cioppi and Marco Pantani would turn over in their graves if they were aware of the above slight. The Forest Service should rectify this glaring abuse of "tradition". (Walt Sniegowski-personal letter in 2012).

First Fire

My heart was racing, blood pressure off the charts, finally headin' for a California fire, not like those piddlin' Arizona brush fires you can pee out, but a raging, death defying inferno racing through the giant pines, trees crashing and me at the head in defiance.

A huge column of heavy smoke filtered thorugh the brush tops, so I couldn't see the actual base of the fire. Down in a cottonwood area, heavy brush drainage with thick white curly smoke, I had to bend my head to see through the undergrowth. The fire crowned and took out a row of dried cottonwood treetops in a flash. The crew boss huddled us, laid out an escape route (first thing you always do), in case things get tense.

Stuck an R5 nozzle in my hands attached to 2700' of hose and said nothing. The R5 is specially designed with both a straight stream function (for more intense fires) and a spraying function(for mopup and water conservation). I ran like hell towards the fire, rarin' to go, fell over a log and the brass head rang my chimes. Got up headed right into the thicket. Crew boss ahead choppin' small brush and small trees. The brush kept catchin' the hose and stopping me and I was ankle deep in water in a small creek and still bent over to try and see (between muffled coughs), smoke thick as New Orleans fog and down to the ground. We broke through the brush and I was hot to attack and a man held out his hand like a traffic cop and yelled "stand clear". Another group of six firefighters swingin' noisy, wide open running chainsaws right and left cut across at right angles. I was left impugned.

The crew roared through and I finally opened my genuine R5 and nothing happened. Not a drop of water. Followed the hose back through the smoky creek, traced the snaky outline back to the truck, got another nozzle, put it on, followed the long snake, back through the water and the smoke, threw her wide open and at the same point in time, about a hundred gallons of water hit me like a sudden storm in summer time. Hadn't even noticed the chopper directly overhead. By the time I recovered the fire was reduced to a few isolated smoke spots and some ashes that I actually got to dribble some water upon. Hell of a start. Between the choppers, the firetrucks the roar of the fire and just bein' busy one tends to become preoccupied. Some giant of an LA county fireman came walkin' by with a hat way too big for his head, slapped me on the back and said, "this your first smoke huh?" I said "yep". Rolled up the hose through the muck and went back to Mill Creek fire station.

Mill Creek Fire Station (aka Tie Summit R. S.) in the late 1800's during construction of the Southern Pacific Railroad, is where railroad ties were cut from logs (predominantly Ponderosa and Jeffrey Pines). Harvested in the area. Much of this work was done by Chinese workers. The names of original work camps/ sidings still remain today. Ravenna, Soledad and Acton to name a few.

Mill Creek was totally destroyed during the 2009 "Station Fire". It was rebuilt and recently rededicated. Most of the timber was destroyed(the original source for the ties) during the 2009 fire. Some of the most intense burning occurred in the Mill Creek area due to a management decision to abandon all suppression efforts in the area.

Four Peak Fire

The fire burned all day on Four Peaks in the Mazatzals Mountains (pronounced Matazels) in central Arizona. Took the ancient government truck with three other forest service firemen up Kline cabin road, where a chopper took us to the ongoing smoke. We were replacing men who had been on the fire all night. The chopper sat down on Lone Pine Ridge. We walked toward the chopper to board and waited first as protocol demanded, for the pilot to give the hand signal to board. We loaded with heads down- tools and hard hats secured.

A trained loader puts the tools up in a special side carrier on the ship. Loaders get on last. Suction from rotor blades which pulls up fire gear into the quickly rotating blades has killed lots of people. Rather, the suction pulls unsecured gear, water cans, axes or shovels into the blades and its Adios Casoose. The ship increased vibrations gathering rpms and took off straight up and fast over the pine covered ridge, heading south.

The wind was over the allowable thirty knots and the ship was bouncing like a rubber ball, headed for the outlines of the four peaks and a small ridge on top. The pilot feathered back and tried to get close, I was closest to the exit side door, saw the landing skid touch down and another firefighter reaching up to help us down. It wasn't much of a landing area.

A sudden blast of wind made the ship shake as I was ready to open the door, the pilot hit the gas pedal and we tore off side saddle dropping straight into the deep canon.

She gained speed and rpms and thank God altitude and we settled for a landing a quarter mile from the fire. We unloaded, checked our bearings, which wasn't needed, because a walking trail led straight to the smoke.

A cold wind was blowing (this was at an altitude of 7,000'), the earlier crew walked past, said they about froze to death in the night. The trail led to an old Amethyst mine (the crystals are hauled out periodically by private chopper). This amethyst, unlike the Brazilian type does have phantoms, but the color can be a very deep purple.

Since the fire was close to the mine, it was named – as they used to be, by a closely connected geographical feature, historic ranch, or sometimes who fought the initial fire. This one is on record as THE AMETHYST FIRE. By the time we arrived at the fire scene, it wasn't. The fire had completed a run up to the tips of the four very vertical peaks which give the mountain range its name, and pooped out. Cold fire, the best kind. Not an ember could be found. Spent the night as warm as our clothing allowed, laying on some backpacked rudimentary sleeping pad and ignoring the insect world beneath our bodies.

Every insect in this part of the world can either stab, fang or zap ya! Beneath sleeping smoke eaters could lie Centi "hundred" Pedis "foot" (Centipedes). They can have from 20 to over 300 pairs of legs. A key trait uniting this group is a pair of "VENOMONOUS CLAWS" (forcipules). Size can range from a few mm's to 12 inches long. There are about 8,000 species worldwide. While they can't stab you, those curved parts at the head can bite the hell out of you. No they're not venomous, but great pain can ensue. Don't forget scorpions, order: Scorpiones, class: Arachnida. They have 8 legs and a segmented body which ends in a very serious stabbing mechanism. The sting is about as painful as a bee's, but their venom can also numb the arm for days at a time. Don't worry about the big ones, those little translucent hummers with a body as big as your thumb can kill a small child. Vinegaroons are weird looking resembling small pets a Martian would like. They're usually no longer than a couple inches in length, many species have scorpion-like pinchers. Lacking venom glands they can spray you with a type of acetic (vinegar-like) acid, hence Vinegarroons They feed mainly on insects and millipedes, but prefer the soft exposed fleshy parts of sleeping firepeople. They are carnivorous and crush their prey between special teeth found on the second segment of their legs (Trochanter). Last, but not least is the invisible insect called a no-seeeum(aka biting midges). These little hummers give no early warning as to their intentions. Sitting by a warm fire with some old cowboy friends in the surreal beauty of an Arizona evening. Coffee warm and strong, some night noises (lonesome coyotes maybe) and without warning your companion jumps straight into the air as if shot and yells unprintable epithets.

The no-see-um has claimed another victim. No one will pass to the great beyond but a savage itch remains for a long time. And the prepetrator is never seen.

Came the morning we scratched around with our pulaskis and shovels trying our best to look busy, continued shoveling around some bushes to be sure all was cold. Since it was the busy time of the fireseason, and the number of helicopters in use on the forest is usually small, we had the not unusual privilege of walking down the entire mountain without benefit of shade or trail, not to mention food or water.

We walked about three miles to the truck, but were more than glad to see it. Walking down hill sounds easier, but the muscles in the front part of the leg are largely unused, so we walked like gorillas with back problems for the rest of the week. The highly identifiable truck stopped at the now defunct Sunflower Cafe, and after a glorious, arterially plugging meal, not to mention some liquid beverages of the alcoholic variety, we returned to the Sunflower Work Center and waited for the next fire call.

Gentle Ben

On Canyon Lake, as the visitors hours numbers increased to well over a million people a year, so did the seriousness of our problems. Years ago when the group pressures were small the work was much more leisurely and personal contact was one of the reasons people enjoyed working for the Forest Service.

Then came the Vietnam War and the death of our cultural shaping forces, the home, the church and the schools. And the hostility became visible between certain age groups and anyone in uniform. Fire prevention and recreation techs titles were changed from techs to "agents." Some even went to law enforcement training with the FBI or CIA. They were supplied weapons, handcuffs and German shepherds.

We used to drive into campgrounds in our truck and people would wave us over for coffee or just to say "hello". When we became, "Green Pigs" the halcyon days were over. This started with having to enforce entrance fees and then the people who loved you would no longer talk - we became tax collectors in uniform with badges and nametags on plastic wood who were now faceless, but also unarmed.Too many people with hostile attitudes on too little land. Instead of writing, "warnings", we wrote'em up and they had to visit the federal magistrate, who often let'em off.

But on the other side, there were those who did not mind spending the dollar, because my partner Bill Blackwell and I explained that most of the money went for improvements to the local campgrounds. The great majority of lake visitors were in this category, just greatful to get out of Dodge City for a while.

Ben, known as Gentle Ben, was a 220 pound bear of a man. He took no lip from anyone. He was a county sheriff and young and we sometimes patrolled together. I walked with Ben down a small trail to the lakeside. Checking young couples - blanket sitting, booze downing, I'D checking.

Our foursome from California (two he's and two she's) seemed benign enough. Except the girls were underage. Ben gave them twenty four hours to leave the state or, "I'll write you up!". So, Ben and I turned to go back up the trail, and I saw Ben whirl around, weapon drawn and heard the word "freeze". Which I did.

Seems the older men had pulled their blankets over weapons unseen by me, but Ben saw the weapons outlined. Two pistols, one a .38 loaded and ready to go, along with a .45 (same condition). So we six formed a long line, the four in front with all their gear, me holding weapons, back to the deputy station. Something about "concealed weapons' now. The four got into their van and we gave them directions to the station. Ben had an extra pistol and handed it to me. "If they come out shooting" - "Blast 'em"

I answered in my best male basso (something like Marshall Dillons' voice) "right, Ben". Ben was assuming that maybe we didn't have all of the artillery. Anyway, they were written up and that was that. Later, in quick reflection, I thought," What if they actually had come out firing?" What if?

Last time I saw Ben was at an ASU football game. Some young drunk had wandered onto the field noticed by Ben. Ben took him down like a wrestler nailing an opponent to the mat. The young man will remember that day as long as he exists.

Reflections

I drove into the Phon D.Sutton campground located at the point where the Verde and Salt rivers meet. Nothing but mesquite trees and thick sand. The river flowed swiftly by and hawks were sitting on the limbs of the cottonwoods. Huge numbers of carplike fish could be seen in the clear areas close to shore. The waterflow isn't regulated by nature anymore, but by man.

Four dams upriver from this point have been built since the early 1900's. The first one being the most distant from this campground, Roosevelt Dam. Italian stone cutters were brought in and cut the stone at the site. Teddy Roosevelt was there for its dedication.

A few of my cowboy friends, now crossed over, were there. Some rode on horseback for two or three days just to hear the presidential speech. Governor George Wiley Paul Hunt, the first state governor of Arizona was there. Being too heavy to ride a horse, the guv arrived in a car.

In years of heavy rain even the four dams filled to capacity could not contain all of the water. So this overflow roars down the same place and the area in this campground still gets flooded for a time. And Phoenix endures the closing of most of her bridges and the people wait until the water recedes.

Phon D. Sutton campground is along the the beautiful Verde River where the young climb onto inner tubes, drink themselves into oblivion, and some of them don't go home. It is also the tubers exit point as a small dam stops everyone just below this point. Often 25,000 souls float this stretch of the river on a typical summertime weekend.

It was in this general area that Ohio James D. Pattie split his forces in the 1840's in a giant recoinnoiter of the land. Looking at 1880 photos, the Salt river was a flowing stream dividing farms and ranches far up the Salt River where Roosevelt dam and townsite now exist. The old large hotel down on the point is gone - now underwater. Old friends the Roscoe Willsons stayed there on occasion.

In historic times, the eagles were numerous, the cottonwoods covered the length of the meandering Rio de Sal and the humpback chubs migrated every

74

year - clear up through smaller creeks, including Cherry and Coon Creeks. A cowboy friend Buss Ellison told me he could ride his horse across Cherry Creek and his horses hooves would be dry, due to the countless numbers of fish. These now vanished chubs covered the entire width of the streams. Eagles were abundant because the fish were their main food and the countless countwood trees provided great nesting places. Even then periodic floods caused much destruction to ranches along the river.

The lower points adjacent to the river have been used from ancient times. The Hohokam Indians in the eighth century built small agricultural communities being the first irrigators in this area. At the end of the civil war, the cowboys rested their cattle here - that great Texas migration - families and cattle - stopped enroute to the high country in the Tonto Basin, Cherry Creek Drainange or the even higher Mogollon Rim. Better grass, fewer people.

The people of Tucson with two million souls in the broader area exist on underground fossil water. Estimates abound as to the abundance of living water lying beneath the desert soil. There are no dams or rivers which supply the water for Tucson.

Centuries earlier in the Phoenix area, those great irrigators, the Hohokam had to abandon their large villages adjoining the Gila River, the amount of water useable for irrigating their crops was greatly reduced through time, and irrigation became impossible. Entire large villages vanished at a time when a thousand people were considered to be heavy impact on the land. We still can't find them after AD1500 - one of the great Southwestern mysteries.

There is a message from the vanished people and their long lost and abandoned canals. It is carried on dry desert winds in the night.

The silver shiny ship headed right for me barely flying above a huge mesa and the gorilla snot (slurry) was coming in at a hundred miles an hour, pushed by two thousand pounds of pink goo. Hit the dirt, belly down, axe grabbed tightly and downslope, other hand on hardhat, ka-whooomp. Left a hole three feet deep close by and split some six foot tall paloverde trees as if they were kindling.

The bomber slurry load (Phos-chek) can rearrange the molecular structure of one's body. Bomber roared and swooshed off, quickly disappearing from view.

The retardant slurry/phos-chek has the chemical name di-ammonium phosphate, which is actually a fertilizer manufactured mainly by Monsanto Chemical Company. It's pink color is a biodegradable dye which assists pilots visually to ascertain how accurate their drops were.

The FS uses many types of aircraft in suppression activities. Included are PBYs'(Flying boats), TBM's (Torpedo bombers), B-17's, P2V's and F7f's. Various helicopters using water slings are also used in ongoing fires.

I got up, rejoined the crew making line down the hill. Being in the lead and workin' my tail off, throwin dirt, shoveling rocks, makin' that line two feet wide to mineral soil. Noticed the quiet. The whole crew behind me had disappeared. I walked back uphill, quite puzzled, and began searching for the lost.

Standing in a circle, bent over, "ooh's, and "looka that's", I joined the formerly working group still standing in a circle, and looked down at the object of interest. A large tarantula, old guy because his belly was nice and fat and those long hairs were really white - was walking tarantula style. Alternating opposite legs two at a time, slow motion, stop, freeze, repeat. Seems most of the boys from east of the big river had never seen this deadly creature This was not what they expected.

We left the arachnid alone, and resumed the line scratching, shovel rock throwing, pulaski chopping grind.

Out of the blue, a few well placed words from some authority figure, who had apparently also noticed the lack of line progress received immediate attention and all of us scrambled to resume the grind. The wind calmed down, the fire was almost out, so we built line all night and by dawn, both the fire and us were ice cold.

The tarantula had no interest in us, I suppose, he, with his terrible near-sightedness, tarantuled off to hunt for supper.

Night Spirits And Bears

Archaic night descends silently, menacingly upon the resting, breathing mountains. There is a darkness on the mountains at night when the moon is non-visible, which leaves the soul with a degree of fear and helplessness. Should there be trouble of any kind, either through deliberate human action or those of the four legged mammals, there Is no one to help on the mountain should trouble of any major type develop. There are large animals here, normally benign, that can cause great bodily harm. Except for extreme, physiological emergencies, I would no more walk down the blackest of trails to the "cabin by the sea", that I would stay on this mountain unarmed. (We were not allowed to have weapons The mountain breathes in geological time. Sleep is deep and refreshing at 8,000'. The smell of pine trees, the whisper of the night wind and the thinness of the air are invigorating. Yet, sometimes, unrecognized sounds gently fill the dark. Neanderthals felt the night evil, the unseen deadly giants of the darkness - lurking subconsciously just outside the fire ring. Here, it is just outside the cabin doors.

Only the creatures of the night move here, sometimes I can hear them. Sometimes I am in control, but on other occasions, the night spirits take over in my mind. Morning brings light and hope and relief. Another night and the Ga-ahn anad Kachinas have done their protective work. The beings of the eternal third world return to middle earth through their sipapus. They have protected me through another time of darkness.

I have on occasion hit a large bear on the butt with my knuckles, almost fractured my metacarpals. Said bear was eating from a garage can and butt pushing my tent. Same week, a mother black bear and her twin cubs met me on the trail in Tennessee, I saw her first walking quickly about 20 feet off the trail. Stood erect-none moving-like the trees. Momma swung her giant head at me, took her cubs up the hill.

Watched the Cibicue Apaches do the "Warrior Dance" once, I was approached well away from the ritual. By luck, a cousin of someone I had taught in my choir at Globe High School (1974-1967) allowed me to watch from a decent distance.

Thanks to my singer, Velasquez Sneezy of San Carlos, Arizona. The dance was a sacred honor for some young man off to war for the U. S. Army. The dancers hidden surrounded by the tall pines. Moved to the sacred four colored directions. Eternal time. Long ritual completed. Silently back to the truck. Leave everyone in peace. Loud talking is considered indecent. Fine.

Smokey The Fire Maker

For decades, the little bear (badly in need of some slimfast) with that ugly park service hat told us all to crush our smokes and douse our campfires -which was good. It is now difficult to explain that the forest service is now in the business of burning everything they can. At the right time, the right wind, the correct humidity, etcetera. Without the annual burning, the new grasses just don't burst forth, the invaded species are killed off, which allows the regrowth of the ancient meadows. The original Americans and the original cattlemen knew this, even though the Texans burned off the original grass and with the incredible running of countless cattle on a range already damaged, created the current situation. In many areas the land has never recovered and cattle are not allowed to graze.

There are other considerations:

Too many acres to manage with too few people. Volunteers, the national guard and even inmates are used during fires.

Mechanical equipment (primarily bulldozers) are destructive to environmentalists, expensive and their use is limited by a rule regarding topography, "no use of dozers with a slope greater than 30%).

Chemical treatment (although very effective) is offensive to environmentally sensitive citizens (especially after Love canal and other superfund sites).

Agent orange, 2-4-5-T usage, Dioxin backlash, and in Arizona, deformed goats and other small mammals due to the use of these chemicals via helicopter delivery. Sixshooter canyon in Globe, Arizona is a prime example. Finally settled after decades of legal wars.

What's left? Prescribed burning release of organisms that will eat the vegetation (i.e. termites, beavers, dermestid bettles (they eat cellulose) or porcupines.

That's leaves only prescribed/prescription burning. So called because of the opinion of some that the land is somehow "sick" and needs a remedy to restore it to health...Synonyms include:

Control burns: Many get out of control.

Management burns: Many show poor management decisions.

Let her burn: A policy that allows in specific areas, and under specific fuel/weather paramenters-certain naturally caused fires(lightning) to burn freely without or with very limited suppression constraints. This is also called a, "natural fire policy" as it mimics what would occur under natural conditions.

Usually fires at higher elevations use the "natural" policy as the fuel is often sparse and very scattered, weather usually more favorable and risk of the fire escaping much lower. Definitely not useable in Southern California with millions of people adjacent or among much more dangerous fuel types than in upper elevations- at the urban-wildland interface.

When I sat as fire lookout on Signal Peak in Arizona, I would get a call and the message would be," The Indians (in this case the Cibicue Apaches) are gonna be burnin' all week. I'd get the coordinates from the fire tower, so I wouldn't report the smoke as something that needed our fire crews. Watched it burn for weeks in the far distance, towering black smoke, miles high.

On an early fire, one long time worker (let's call him Les) was told to take a fusee (30 minute type - about three feet long), and slowly walk down the meandering road which would give everyone time to slowly stamp out the small flames, which Les was creating.

Well, Les was close to retirement, and with a gleam in his eye, as soon as he could, Les took off like a man possessed, first at a slow, ambling gait and then faster and faster down the road. Mother nature helped by bringing in a monstrous wind - blowing in from nowhere.

This fire roared through turbinella oak so thick that a snake had trouble getting through. The fire stopped in the far distance, when it ran itself out on some forlorn high rocky ridge - which simply meant there wasn't any fuel left to burn. This was a "good" fire. This was a "great" fire.

Because of the control burn, all of the invader species were burned out and the grasses and native brush could return. In many places, formerly open to even wagon teams in the early, pioneer times, the invading growth had closed the trails to even hikers.

Burning, in some cases, reopens old trails, and best of all, in one area I know of, known locally as Tejanos Spring, after a huge control burn, several small springs returned, bubbling up once again. Old -timers in the area could not recall in their memory, when the last springs were visible.

The only thing remaining, that has to go is the four footed, thick-skulled introduced bovines. When they are gone, and the burning continues, then the land will slowly heal. That's my opinion. Other specialists believe that if the numbers of said critters was reduced and well managed, they might replace the long ago grazing mammals as bison, pronghorn, bighorn(originally a plains species), elk and deer.

Tribute To George

On the meandering road from Pasadena towards Palmdale, California, George was a tanker foreman in a station in one of the small canyons in an area known as Hidden Valley.

About 6'3", thick, dark moustache, never met a nicer man. In 1977, he received a fire call in the off season in the middle of the night and roared out of the station headed down canyon not knowing he would never return. Somewhere between Monte Vista Fire Station and the fire, a raging wall of water had removed the road and destroyed a small tavern where the Mill Creek Tanker Crew used to eat in the evening. I heard many people were killed. This same wall of water grabbed the giant firetruck and hurled it into eternity, the green door torn off the truck was found a 1/2 mile from where he drove into it, his body was found weeks later.

Another crewman from Monte Vista survived with his feet on the body of someone else, the brush wrapping around him, and an isolated tree stopped his movement, and preserved his life. We owe something to the George's who die too young. Of course, to me he's not gone, I can close my eyes and see him, even now. I heard later that his son has followed in his fire fighting tradition. (George is seen in the back row of the photo of the Mill Creek Firecrew).

Disastrous fires through historical time

	Size	Name	Area	Notes
1846				
1853	0,450,000 450,000 acres (180,000 ha)	Yachina Fire	Oregon	
1868	0,320,000 320,000 acres (130,000 ha)	Nestucca Fire	Oregon	
1870	0,300,000 300,000 acres (120,000 ha)	Coos Fire	Oregon	
1876	0,964,000 964,000 acres (390,000 ha) [2]	Saguenay Fire[3][4]	Quebec	
1889	0,500,000 500,000 acres (200,000 ha)	Bighorn Fire	Wyoming	
1903	0,300,000 300,000 acres (120,000 ha)	Santiago Canyon Fire of 1889	California	
1948	0,464,000 464,000 acres (188,000 ha)	Adirondack Fire	New York	
1950	0,645,000 645,000 acres (261,000 ha)	Mississagi/ Chapleau fire	Ontario	
1951	0,017,000 17,000 acres (6,900 ha)	Capitan Gap fire	New Mexico	
1951	0,048,052 48,052 acres (19,446 ha)	McKnight Fire	New Mexico	
1995	0,038,000 380,000 acres (150,000 ha)	Great Forks Fire	Washington	
2005	0,007,000 7,000 acres (2,800 ha)	Sunrise Fire of 1995	New York	
2007	0,017,000 17,000 acres (6,900 ha)	September 2005 California wildfires	California	
2012	0,653,100 653,100 acres (264,300 ha)	Murphy Complex Fire	Idaho - Nevada	
2012	0,248,000 248,000 acres (100,000 ha)	Ashland Fire	Montana	
2012	0,075,431 75,431 acres (30,526 ha)	Chips Fire	California	
2013	0,332,000 332,000 acres (134,000 ha)	Mustang Complex Wildfire	Idaho	
2013	0,018,800 18,800 acres (7,600 ha)	Silver Fire	New Mexico	0% contained as of 6/12/2013 [14]
2006	0,003,538 3,538 acres (1,432 ha)	Jaroso Fire	New Mexico	0% contained as of 6/13/2013 [15]

1846	Size	Name	Area	Notes
2013	0,040,200 40,200 acres (16,300 ha)	Esperanza Fire	California	10 buildings destroyed, 5 firefighters killed. The blaze started on October 26 and scorched 40,200 acres (16,300 ha), or more than 60 square miles (160 km²), of forest and brush before being fully contained October 30. It destroyed 34 homes and 20 outbuildings.
2013	0,003,218 3,218 acres (1,302 ha)[19]	Royal Gorge Fire	Colorado	100% contained as of 6/16/2013;[19] Jumped Royal Gorge and damaged the Royal Gorge Bridge.
2013	0,014,198 14,198 acres (5,746 ha)[16]	Black Forest Fire	Colorado	100% contained as of 6/21/2013; Large, fast-spreading fire due to dry conditions, high heat and restless winds. The 14,280 acre fire has destroyed 509 homes and left 17 homes partially damaged. As of June 13, 2013 it became the most destructive fire in Colorado state history. ATF and state officials are investigating the point of origin and cause of the blaze that claimed the lives of two people.[17] As of June 21, 2013 estimates of damage are expected to exceed $90 million.[18]
2002	0,020,000 20,000 acres (8,100 ha)[22]	Mount Charleston Fire	Nevada	15% contained as of 7/9/2013
2013	0,499,570 499,750 acres (202,240 ha)	Florence/Sour Biscuit Complex Fire	Oregon	150 million dollars to suppress.
2013	0,001,300 1,300 acres (530 ha)[20]	Yarnell Hill Fire	Arizona	19 firefighters killed on June 30, 2013.

1846	Size	Name	Area	Notes
1998	0,126,000 126,000 acres (51,000 ha)[25]	Beaver Creek Fire	Idaho	2,300+ homes evacuated 7,700 homes are under pre-evacuation. The fire is 9% contained as of the morning hours of August 18, 2013.
2013	0,300,000 300,000 acres (120,000 ha)	Unnamed	Florida	2200 fires, during drought season; burned 150 homes, $390 million timber lost, 80,000 evacuees, $133 million in fire suppression costs
2013	0,025,000 25,000 acres (10,000 ha)[23]	Bison Fire	Nevada	25% contained as of 7/9/2013
2007	0,018,000 18,000 acres (7,300 ha)[24]	Silver Fire	California	26 homes burned and more than 500 threatened. Charred 30 square miles in three days and 70% as of August 10, 2013.
1970	0,003,500 3,500 acres (1,400 ha)	Angora Fire	California	3 injuries.
2002	0,175,425 175,425 acres (70,992 ha)	Laguna Fire	California	382 homes destroyed and 8 people killed; the largest fire in the state's history until the Marble Cone Fire
2012	0,137,760 137,760 acres (55,750 ha)	Hayman Fire in Pike National Forest	Colorado	5 firefighter deaths, 600 structures fires
2013	0,028,098 28,098 acres (11,371 ha)	Ponderosa Fire	California	52 residences & 81 outbuildings destroyed (131 total); 1 residence & 5 outbuildings damaged
2013	0,023,946 23,946 acres (9,691 ha)	Thompson Ridge Fire	New Mexico	60% contained as of 6/13/2013 [13]
2013	0,010,282 10,282 acres (4,161 ha)	Tres Lagunas fire	New Mexico	80% contained as of 6/13/2013 [12]

1846	Size	Name	Area	Notes
2003	0,024,251 24,251 acres (9,814 ha)[9]	Springs Wildfire	California	95% contained as of 5/8/2013; Unusually large fire for springtime. Despite burning near residential areas, no homes were damaged. [10] (Name comes from the location of the fire - Camarillo Springs Rd)[11]
2013	0,091,281 91,281 acres (36,940 ha)	Old Fire	California	993 homes destroyed, 6 deaths. Simultaneous with the Cedar Fire.
1947	0,007,000 7,000 acres (2,800 ha)[26]	Little Queens Fire	Idaho	A mandatory evacuation order was issued for Atlanta, Idaho residents on August 20, 2013. Believed to be human caused.
2007	0,175,000 175,000 acres (71,000 ha)	The Great Fires of 1947	Maine	A series of fires that lasted ten days; 16 people killed
2009	0,127,244 127,244 acres (51,494 ha)	California wildfires of October 2007	California	A series of wildfires that killed 9 people and injured 85 (including 61 firefighters). Burned at least 1,500 homes from the Santa Barbara County to the U.S.–Mexico border. Aggravated by Santa Ana winds that reached up to 85 mph (140 km/h). The largest fire, the Witch (Creek), was located in San Diego county.
2013	0,157,220 157,220 acres (63,620 ha)	Station Fire	California	As of 9:51am PDT September 6, 2009; The Station Wildfire burned 157,220 acres (636.2 km²), and is currently the 10th largest in state history. There were 4,735 personnel assisting in the firefighting efforts.[7][8]

1846	Size	Name	Area	Notes
1986	0,222,777 253,332 acres (102,520 ha) [27]	Rim Fire	California	As of September 7, the fire was 80% contained and the third largest fire in California history. Over 100 buildings have been reported damaged. 9 injuries have been reported.
2000	0,073,000 73,000 acres (30,000 ha)	Topsail / Holly Shelter Fire	North Carolina	Burned 80 percent of the Holly Shelter Game Lands and sent smoke wafting over Wilmington; cost $308,000 to contain [5]
2008	0,048,000 48,000 acres (19,000 ha)	Cerro Grande Fire	New Mexico	Burned about 420 dwellings in Los Alamos, New Mexico, damaged >100 buildings at Los Alamos National Laboratory; $1 billion damage, second worst fire in state's recorded history
2003	0,013,709 13,709 acres (5,548 ha)	Trigo Fire	New Mexico	Burned from 15 April to 22 May. 59 homes were destroyed. The fire had a containment cost of $11 million.
2003	0,084,750 84,750 acres (34,300 ha)	Aspen Fire	Arizona	Destroyed large portions of Summerhaven, Arizona
2013	0,061,776 61,776 acres (25,000 ha)	Okanagan Mountain Park Fire	British Columbia	Displaced 45,000 inhabitants, destroyed 239 homes and threatened urbanized sections of Kelowna.
2000	0,222,777 7,400 acres (3,000 ha) [30]	Clover Fire	California	Fire reported on September 9, 2013 in Happy Valley was burning its way south toward Tehama County. At least 30 structures have been destroyed and at least 350 are threatened on September 10.

1846	Size	Name	Area	Notes
1985	0,079,244 79,244 acres (32,069 ha)	Manter Fire	California	Firefighters were limited to the use of hand tools and aerial support in fighting this fire due to the fire occurring in the Domeland Wilderness Area.
2008	0,093,000 93,000 acres (38,000 ha)	Allen Fire	North Carolina	In 1985, nearly 93,000 acres of forest, wetlands and farmland burned in northeastern North Carolina in one of the biggest fires in modern state history[5]
1871	1,557,293 1,557,293 acres (630,214 ha)	Summer 2008 California wildfires	California	In Northern California, the fires were mostly started by lightning. In Santa Barbara (Southern California), the Gap fire endangered homes and·lives. The Basin Complex and Gap fire were the highest priority fires in the state at this time. Caused unhealthy air quality in large parts of California for several weeks. Near Yosemite the Telegraph Fire was started by target shooters. During all fires many homes were lost.
1949	2,500,000 2,500,000 acres (1,000,000 ha)	The Great Michigan Fire	Michigan	It was overshadowed by the Great Chicago Fire that occurred on the same day.
1994	0,004,500 4,500 acres (1,800 ha)	Mann Gulch fire	Montana	Killed 13 firefighters
1953	0,002,115 2,115 acres (856 ha)	South Canyon fire	Colorado	Killed 14 firefighters
1825	0,001,300 1,300 acres (530 ha)	Rattlesnake Fire	California	Killed 15 firefighters. Well known textbook case used to train firefighters.
1881	3,000,000 3,000,000 acres (1,200,000 ha)	Miramichi Fire	New Brunswick	Killed 160 people.

	Size	Name	Area	Notes
1846				
1916	1,000,000 1,000,000 acres (400,000 ha)	Thumb Fire	Michigan	Killed 200+ people
1991	0,500,000 500,000 acres (200,000 ha)	Great Matheson Fire	Ontario	Killed 228 (U.O. 400+) people and destroyed several towns, Cochrane burnt again after just five years.
2001	0,001,520 1,520 acres (620 ha)	Oakland Hills firestorm	California	Killed 25 and destroyed 3469 homes and apartments within the cities of Oakland and Berkeley
1894	0,009,300 9,300 acres (3,800 ha)	Thirty Mile Fire	Washington	Killed 4 firefighters
1922	0,160,000 160,000 acres (65,000 ha)	Hinckley Fire	Minnesota	Killed 418 people and destroyed 12 towns
1911	0,415,000 415,000 acres (168,000 ha)	Great Fire of 1922	Ontario	Killed 43 people and burnt through 18 townships in the Timiskaming District
1910	0,500,000 500,000 acres (200,000 ha)	Great Porcupine Fire	Ontario	Killed 73 people
1918	3,000,000 3,000,000 acres (1,200,000 ha)	Great Fire of 1910	Idaho-Montana-Washington	Killed 86 people, including 78 firefighters
1871	0,100,000 100,000 acres (40,000 ha)	Cloquet Fire	Minnesota-Wisconsin	Killed between 400 and 500 people
2003	1,200,000 1,200,000 acres (490,000 ha)	Peshtigo Fire	Wisconsin	Killed over 1,700 people and has distinction of the conflagration that caused the most deaths by fire in United States history. It was overshadowed by the Great Chicago Fire that occurred on the same day.
2011	0,090,769 90,769 acres (36,733 ha)	B&B Complex Fires	Oregon	Large fire in Central Oregon between Black Butte and Mount Jefferson. The fire closed off a large section of state HWY 20. The fire began as two separate fires. Both started on August 19 and lasted until September 5.

1846	Size	Name	Area	Notes
2002	0,156,293 156,293 acres (63,250 ha)	Las Conchas Fire	New Mexico	Largest fire in New Mexico state history. 63 homes lost. Threatened Los Alamos National Laboratory.
2007	0,150,700 150,700 acres (61,000 ha)	McNally Fire	California	Largest fire in Sequoia NF history.
2007	0,124,584 124,584 acres (50,417 ha)	Florida Bugaboo Fire	Florida	Largest fire on record in Florida.
1950	0,363,052 363,052 acres (146,922 ha)	Milford Flat Fire	Utah	Largest fire on record in Utah.
2003	3,500,000 3,500,000 acres (1,400,000 ha)	Chinchaga fire	British Columbia and Alberta	Largest North American fire on record
2007	0,280,278 280,278 acres (113,424 ha)	Cedar Fire	California	Largest recorded fire in California history (see 1889 Santiago Canyon fire that may have been larger); burned 2,232 homes and killed 15 in San Diego County. Simultaneous with 15 other fires in Southern California (including the Old Fire) covering 721,791 acres (292,098.5 ha), killing 24, displacing 120,000 and destroying 3,640 homes. Damage from combined fires estimated at 2 billion USD
2004	0,468,938 468,938 acres (189,772 ha)	Sweat Farm Road/Big Turnaround Complex Fire	Georgia	Largest recorded fire in Georgia history. 26 structures were lost.
2012	1,305,592 1,305,592 acres (528,354 ha)	Taylor Complex Fire	Alaska	Largest wildfire by acreage of 1997-2007 time period

1846	Size	Name	Area	Notes
1977	0,289,478 289,478 acres (117,148 ha)	Whitewater-Baldy Complex Fire	New Mexico	Largest wildfire in New Mexico state history. Began in the Gila Wilderness as two separate fires that converged, both started by lightning. Destroyed 12 homes in Willow Creek, NM.
2012	0,178,000 178,000 acres (72,000 ha)	Marble-Cone Fire	California	Lightning caused at end of La Niña drought, burns Ventana Wilderness in Big Sur area; the largest fire in recorded state history until the Cedar Fire
2013	0,018,247 18,247 acres (7,384 ha)	Waldo Canyon Fire	Colorado	Located near Pikes Peak, north and west of Colorado Springs in the Waldo Canyon - origin currently unknown - first reported the afternoon of Saturday, June 23. Destroyed 346 homes making it the second most destructive fire in state history. Two fatalities reported.
2012	0,025,000 3,718 acres (1,505 ha)[28] [29]	Morgan Fire	California	Mandatory evacuations for Oak Hill Lane, Curry Canyon and Curry Point in Morgan Territory. 50-75 structures threatened and 20% contained as of September 9, 2013.
1988	0,044,330 44,330 acres (17,940 ha)	Little Bear Fire	New Mexico	Most destructive wildfire in New Mexico state history. Began in the Lincoln National Forest and was started by lightning.
2009	0,793,880 793,880 acres (321,270 ha)	Yellowstone fires of 1988	Wyoming-Montana	Never controlled by firefighters; only burned out when a snowstorm hit.

1846	Size	Name	Area	Notes
2011	0,024,406 24,406 acres (9,877 ha)	West Kelowna Wildfires	British Columbia	On July 18, 2009, 3 wildfires started within hours of each other in and around the city of West Kelowna, which burned out of control until August 20. (Terrace Mountain Fire, 9277 hectares) (Glenrosa Fire 400 hectares) (Rose Valley Fire, 200 hectares) 25000 people were evacuated and 4 homes were burned during the first day of the Glenrosa Fire.
2012	0,012,000 12,000 acres (4,900 ha)	Slave Lake Wildfire	Alberta	On May 14, a wildfire deliberately set 15 kilometres (9.3 mi) west of Slave Lake moved eastward towards the town, fueled by 100 kilometres (62 mi) winds, burned until May 16. Nearly 7,000 people were evacuated from the town and nearby surrounding communities. 374 properties were destroyed and 52 damaged in Slave Lake while an additional 59 properties were destroyed and 32 damaged in the surrounding communities. The only fatality was that of a pilot who died when his helicopter crashed while battling the fires. The overall damage cost was $1.8 billion.
2013	0,719,694 719,694 acres (291,250 ha)	Long Draw Fire and Miller Homestead Fire	Oregon	Oregon's largest fire in 150 years.
2008	0,617,763 617,763 acres (250,000 ha)[21]	Quebec Fire	Quebec	Over 300 evacuated.

1846	Size	Name	Area	Notes
2012	0,041,534 41,534 acres (16,808 ha)	Evans Road Wildfire	Eastern North Carolina	Peat fire started on June 1 by lighting strike during North Carolina's drought - the worst on record. 450 firefighters battled it. 71 high capacity pumps move billions of gallons of water. It burned for three months.[6]
2012	0,023,500 23,500 acres (9,500 ha)	Taylor Bridge Fire	Washington	Started Aug 13, 2012, fully contained on Aug. 28. At least 170 structures lost including 61 homes. 23,500 acres burned (36.7 sq mi).
2011	0,087,284 87,284 acres (35,323 ha)	High Park Fire	Colorado	Started by lightning, it is the Second Largest wildfire in Colorado state history by size. It killed one person and destroyed at least 248 homes making it the most destructive fire in state history until Waldo Canyon Fire a few days later.
1933	1,748,636 1,748,636 acres (707,648 ha)	Richardson Backcountry Fire	Alberta	Started early in the spring of 2011, as of October 2011 the fire was over 700,000 Ha in size and still burning.
1939	0,240,000 240,000 acres (97,000 ha)	Tillamook Burn	Oregon	Swept through the same region of Oregon four times
1945	0,190,000 190,000 acres (77,000 ha)	Tillamook Burn	Oregon	Swept through the same region of Oregon four times
1951	0,180,000 180,000 acres (73,000 ha)	Tillamook Burn	Oregon	Swept through the same region of Oregon four times
2007	0,032,700 32,700 acres (13,200 ha)	Tillamook Burn	Oregon	Swept through the same region of Oregon four times

1846	Size	Name	Area	Notes
2011	0,240,207 240,207 acres (97,208 ha)	Zaca Fire	California	The blaze was started July 4 by sparks from water pipe repair equipment. The fire had a containment cost of $117 million. It was contained on September 2. It is California's second largest recorded fire.
2011	0,538,049 538,049 acres (217,741 ha)	Wallow Fire	Arizona & New Mexico	The largest fire in Arizona state history. In one 24-hour burn period (6/6-6/7), it consumed 77769 acres of forest land.
1987	0,034,000 34,000 acres (14,000 ha)	Bastrop County Complex fire	Texas	The worst fire in Texas state history,destroyed over 1500 homes
2002	0,650,000 650,000 acres (260,000 ha)	Siege of 1987	California-Oregon	These fires were started by a large lightning storm in late August. The storm started roughly 1600 new fires, most caused by dry lightning. Firefighting efforts continued into October, before the majority of the fires were controlled.
1868	0,467,066 467,066 acres (189,015 ha)	Rodeo-Chediski fire	Arizona	Threatened, but did not burn the town of Show Low, Arizona
	1,000,000 1,000,000 acres (400,000 ha)	Silverton Fire	Oregon	Worst recorded fire in state's history

Index To The Wildfire Portion Of The Work

The author may be reached by the following methods:

RESUME OF PHILIP G. SMITH

1960 - Education B.A., Asbury College. Church Music Major. M.M.E., North Texas State University.
(1964 – 1969 – Choral director at high schools in Globe, Arizona, Pueblo, Colorado,Artesia, New Mexico), Wickenburg, Arizona. (1974 – 1976) and Union High School (1982 – 1983). Director for Rotary men-on-note in Tulsa, Oklahoma
1972 - Scottsdale Community College, Scottsdale, Az. Two string pieces premiered by the Phoenix Symphony. Paramedic for Mesa Ambulance.
1972 to 1974 Firefighter - Fire suppression in Arizona and California.
1980 – Field archeologist , University of Arizona excavation of numerous prehistoric sites in Arizona. Tucson Opera Chorus. Draftsman for C. E. Natco, Tulsa, Oklahoma.
1985 - 1988 Orme School, Mayer, Arizona, Chairman of Fine Arts. Instructor in music history, archeology, choir, archeology caravan and head wrestling coach.
1988 - 1989 Independence, kansas (wife's home)family illness,accompanist for musical "Mame" which played to 2,000 people in three days.
1989 - 1992 Kiski Prep, Saltsburg, Pennsylvania, Director of Mens' Glee Club, Western Civilization instructor and coached golf, wrestling. Dormmaster.

1993 – 2004 Camp Verde Public Schools. Music and Archeology teacher. (Excavation of the Swetnam Fort (1865 – 1867) and Camp Lincoln (1865 – 1871). Fire Tower Lookout on Mingus Mountain (2001 to 2003).

2004 – 2013 Claremore, Oklahoma. Music composing. Over 300 pieces for all media. Retired as Organist/Choirmaster at St. Paul's Episcopal Church. Assisted the University of Oklahoma in field excavations at a Wichita Indian site on the Arkansas River (1750's). Excavated at Fort Gibson (when Sam Houston was there) and helped recover a mammoth skeleton in Reno, Oklahoma.

Philip G. Smith
P. O. Box 1247
Claremore, Oklahoma 74018

918 – 342 – 3536.

Email: philip0284@sbcglobal.net+

A Word to the gentle reader

Upon my retirement from teaching school, I was searching through some dusty, old, cardboard boxes.

The nucleus of what has become this book was within. This look at the forest service reflects my thinking from the early 1970's, the time period in which I was a fire fighter.

Fire fighting techniques, tools, training and philosophy have all changed since then. But, the great majority of the work is still delegated to the "grunt" on the fire line (much like the army).

I have tried to present an all-round picture of the fire fighter (USFS) during this time period.

Any slur, slander or insult to any agency, animal, group or individual was totally not intended.

I wrote in truly sarcastic tones at times (when needed), and tried to bring a little levity at the same time.

In reflection, the best time of my life was spent eating smoke, walking down trailless mountains and marveling at the great outdoors that the great God has created for us.

I feel sadness for the people of the city who have never been out in the wilderness. They have missed a great part of what it means to be alive.